SpringerBriefs in Earth System Sciences

Series Editors

Gerrit Lohmann
Jorge Rabassa
Justus Notholt
Lawrence A. Mysak
Vikram Unnithan

For further volumes:
http://www.springer.com/series/10032

V. Balaji · René Redler
Reinhard Budich

Earth System Modelling – Volume 4

IO and Postprocessing

 Springer

V. Balaji
Geophysical Fluid Dynamics Laboratory
Modeling Systems Group
Princeton, NJ
USA

René Redler
Reinhard Budich
Max-Planck-Institut für Meteorologie
Hamburg
Germany

ISSN 2191-589X   ISSN 2191-5903   (electronic)
ISBN 978-3-642-36463-1   ISBN 978-3-642-36464-8   (eBook)
DOI 10.1007/978-3-642-36464-8
Springer Heidelberg New York Dordrecht London

Library of Congress Control Number: 2011938123

Printed on acid-free paper

Springer is part of Springer Science+Business Media (www.springer.com)

We thank Panos Adamidis, Jeff Durachta, Brian Gross, John Helly, and Scott Klasky for insightful comments on earlier versions of this manuscript

Preface

Climate modelling in former times mostly covered the physical processes in the Earth's atmosphere. Nowadays, there is a general agreement that not only physical, but also chemical, biological and, in the near future, economical and sociological—the so-called anthropogenic—processes have to be taken into account on the way towards comprehensive Earth system models. Furthermore, these models include the oceans, the land surfaces and, so far to a lesser extent, the Earth's mantle. Between all these components feedback processes have to be described and simulated.

Today, a hierarchy of models exist for "Earth system modelling". The spectrum reaches from conceptual models—back of the envelope calculations—over box-, process- or column-models, further to Earth system models of intermediate complexity and finally to comprehensive global circulation models of high resolution in space and time. Since the underlying mathematical equations in most cases do not have an analytical solution, they have to be solved numerically. This is only possible by applying sophisticated software tools, which increase in complexity from the simple to the more comprehensive models.

With this series of briefs on "Earth System Modelling" at hand we focus on Earth system models of high complexity. These models need to be designed, assembled, executed, evaluated and described, both in the processes they depict as well as in the results the experiments carried out with them produce. These models are conceptually assembled in a hierarchy of sub-models, where process models are linked together to form one component of the Earth system (Atmosphere, Ocean, ...), and these components are then coupled together to Earth system models in different levels of completeness. The software packages of many process models comprise a few to many thousand lines of code, which results in a high complexity of the task to develop, optimise, maintain and apply these packages, when assembled to more or less complete Earth system models.

Running these models is an expensive business. Due to their complexity and the requirements with respect to the ratios of resolution versus extent in time and space, most of these models can only be executed on high performance computers, commonly called supercomputers. Even on today's supercomputers, typical model experiments take months to conclude. This makes it highly attractive to increase the efficiency of the codes. On the other hand the lifetime of the codes exceeds the typical lifetime of computing systems and architectures roughly by a factor of three. This means that the codes need not only be portable, but also constantly adapted to emerging computing technology. While in former times computing power of single processors—and that of clustered computers—was resulting mainly from increasing clock speeds of the CPUs, today's increases are only exploitable when the application programmer can make best use of the increasing parallelism off-core, on-core and in threads per core. This adds complexity to areas like IO performance, communication between cores or load balancing to the assignment at hand.

All these requirements put high demands on the programmers to apply software development techniques to the code, making it readable, flexible, well structured, portable and reusable, but most of all capable in terms of performance. Fortunately, these requirements match very well an observation from many research centres: due to the typical structure of the staff of the research centres, code development oftentimes has to be done by scientific experts, who typically are not computing or software development experts. Therefore, the code they deliver needs a certain amount of quality control to assure fulfilment of the requirements mentioned above. This quality assurance has to be carried out by staff with profound knowledge and experience in scientific software development and a mixed background from computing and science.

Since such experts are rare, an approach to ensure high code quality is the introduction of common software infrastructures or frameworks. These entities attempt to deal with the problem by providing certain standards in terms of coding and interfaces, data formats and source management structures, that enable the code developers as much as the experimenters to deal with their Earth system models in a well acquainted, efficient way. The frameworks foster the exchange of codes between research institutions, the model inter-comparison projects so valuable for model development and the flexibility of the scientists when moving from one institution to another, which is commonplace behaviour these days.

With an increasing awareness about the complexity of these various aspects, scientific programming has emerged as a rather new discipline in the field of Earth system modelling. At the same time, new journals are launched providing platforms to exchange new ideas and concepts in this field. Till date, we are not aware of any textbook addressing this field, tailored to the specific problems the researcher is confronted with. To start a first initiative in this direction, we have compiled a series of six volumes, each dedicated to a specific topic the researcher is confronted with when approaching "Earth System Modelling":

Volume 1: Recent Developments and Projects
Volume 2: Algorithms, Code Infrastructure and Optimisation
Volume 3: Coupling Software and Strategies

Volume 4: IO and Postprocessing
Volume 5: Tools for Configuring, Building and Running Models
Volume 6: ESM Data Archives in the Times of the Grid

This series aims at bridging the gap between IT solutions and Earth system science. The topics covered provide insight into state-of-the-art software solutions and in particular address coupling software and strategies in regional and global models, coupling infrastructure and data management, strategies and tools for pre- and post-processing and techniques to improve the model performance.

Volume 1 familiarises the reader with the general frameworks and different approaches for assembling Earth system models. Volume 2 highlights major aspects of design issues that are related to the software development, its maintenance and performance. Volume 3 describes different technical attempts from the software point of view to solve the coupled problem. Once the coupled model is running, data are produced and postprocessed. Volume 4 at hand touches upon key aspects of this part of the workflow. The whole process of creating the software, running the model and processing the output is assembled into a workflow (Volume 5). Volume 6 describes co-ordinated approaches to archive and retrieve data.

Hamburg, December 2012   René Redler
                                                                                         Reinhard Budich

Contents

Contributors

V. Balaji Princeton University, New Jersey, USA, e-mail: balaji@princeton.edu

John Caron University Corporation for Atmospheric Research, Boulder, USA, e-mail: caron@unidata.ucar.edu

Robert Drach Program for Climate Model Diagnosis and Intercomparison, Lawrence Livermore National Laboratory, Livermore, USA, e-mail: drach1@llnl.gov

Steven C. Hankin Pacific Marine Environmental Laboratory, Seattle, USA, e-mail: steven.c.hankin@noaa.gov

Robert Latham Argonne National Laboratory, Lemont, USA, e-mail: robl@mcs.anl.gov

Don Middleton National Center for Atmospheric Research, Boulder, USA, e-mail: don@ucar.edu

Thomas J. Phillips Program for Climate Model Diagnosis and Intercomparison, Lawrence Livermore National Laboratory, Livermore, USA, e-mail: phillips14@llnl.gov

Robert Ross Argonne National Laboratory, Lemont, USA, e-mail: rross@mcs.anl.gov

Bernie Siebers Geophysical Fluid Dynamics Laboratory, Princeton, USA, e-mail: Bernie.Siebers@noaa.gov

Dean N. Williams Program for Climate Model Diagnosis and Intercomparison, Lawrence Livermore National Laboratory, Livermore, USA, e-mail: williams13@llnl.gov

Chapter 1
Input/Output and Postprocessing

V. Balaji

"I/O certainly has been lagging in the last decade."—Seymour Cray, Public Lecture (1976).
"Also, I/O needs a lot of work."—David Kuck, Keynote Address, 15th Annual Symposium on Computer Architecture (1988).
"I/O has been the orphan of computer architecture."—Hennessy and Patterson, Computer Architecture - A Quantitative Approach. 2nd Ed. (1996).

Quotes from Mashey (1999): http://www.usenix.org/event/usenix99/invited_talks/mashey.pdf.

The problem of reading and writing data from media such as disk and tape while performing computation has long been a challenging one. As the quotes above indicate, I/O has often been an afterthought, a sideshow to the theater of more exciting developments in the field of high-performance computing.

Earth System models, in particular, have always struggled with the I/O burden. In recent years, the problems have become particularly acute, and has focused attention on this critical area. The reasons for the I/O crisis are manifold:

- Earth System Model (ESM) performance at fixed problem size (i.e., time to solution at a given model resolution) has grown only slowly in the era of commodity hardware, forcing the community toward higher resolution models. Since data volume grows as N^3 for problem size N, I/O problems have become relatively more pronounced.
- Disk access rates have not kept pace with processor speed increases over the last two decades.
- Cluster hardware has given rise to parallel and shared filesystems whose performance has not matched the performance of scalable computation. In particular the number of I/O channels into shared filesystems does not match the processor count.

V. Balaji
Princeton University, Princeton, USA
e-mail: balaji@princeton.edu

V. Balaji et al., *Earth System Modelling – Volume 4*, SpringerBriefs in Earth System Sciences, DOI: 10.1007/978-3-642-36464-8_1, © The Author(s) 2013

- Finally, ESM methodology has increasingly pushed toward large coordinated international modeling campaigns, increasing our dependence on shared standards, avoiding hardware-specific I/O optimizations, and toward large distributed data archives.

It is encouraging, therefore, to report that the sense of crisis and the surge of interest in this area seems to be yielding dividends. Progress has been made on basic parallel I/O, on the development of abstract I/O APIs and data models that can be optimized on varied hardware, in storage management and analysis subsystems. This chapter offers an overview of this rich terrain.

In Chap. 2, Latham and Ross cover some basics of parallel I/O, covering parallel filesystems, describing now-standard parallel I/O libraries such as MPI-IO, and high-level libraries providing data layers wrapping MPI-IO and connecting it to community-standard data formats such as NetCDF and HDF. In Chap. 3, Balaji walks us through I/O as practised in ESMs, how parallel I/O is implemented in models, and the supply chain of data from a running model to post-processing to distribution. In Chap. 4, Siebers and Balaji describe the layers of hierarchical storage management, with the largest data volumes stored on linear media (tape) connected to random-access (disk) storage media smaller in volume but closer to the application. This section also outlines a three-level storage abstraction to capture this complexity at the application level. In Chap. 5, Drach and Caron describe the data representations underlying the most commonly used data formats in the field, leading us to a *common data model* underpinning format-neutral access layers for large data archives. Finally, in Chap. 6, Williams, Hankin, Phillips and Middleton describe the data and I/O problem from the perspective of the end user, describing the ESM data analysis process and the community tools and environments that enable it.

Chapter 2
Parallel I/O Basics

Robert Latham and Robert Ross

As Earth systems simulations grow in sophistication and complexity, developers need to be concerned not only about computational constraints but also about storage resources. When dealing with the large data sets produced by high-resolution simulations, the storage subsystem must have both the capacity to store the data and the capability to access that data efficiently. The computing facilities of today and tomorrow will provide increasing computational power, but storage capabilities are not increasing at a corresponding rate. One challenge for all applications will be how to best manage and mitigate the growing input/output (I/O) bottleneck.

By understanding the structure of high-performance storage systems, scientists can better direct their efforts as they strive to get optimum performance out of storage. High-performance storage systems rely on an entire stack of software tools and libraries (Fig. 2.1). This chapter provides an overview of the software stack, discusses tuning strategies for applications, and briefly covers some best practices for high-performance parallel I/O.

2.1 The I/O Software Stack

Each year, high-end computing systems offer ever more powerful hardware. Computational scientists in turn refine software models of physical processes to take advantage of faster machines, doing so with the help of libraries optimized for specific platforms. Similarly, computational scientists can rely on I/O abstraction libraries to both hide platform-specific aspects and deliver higher performance. This organization of I/O software is commonly referred to as the *I/O software stack*. Using

R. Latham · R. Ross
Argonne National Laboratory, Lemont, USA
e-mail: robl@mcs.anl.gov

R. Ross
e-mail: rross@mcs.anl.gov

V. Balaji et al., *Earth System Modelling – Volume 4*, SpringerBriefs in Earth System Sciences, DOI: 10.1007/978-3-642-36464-8_2, © The Author(s) 2013

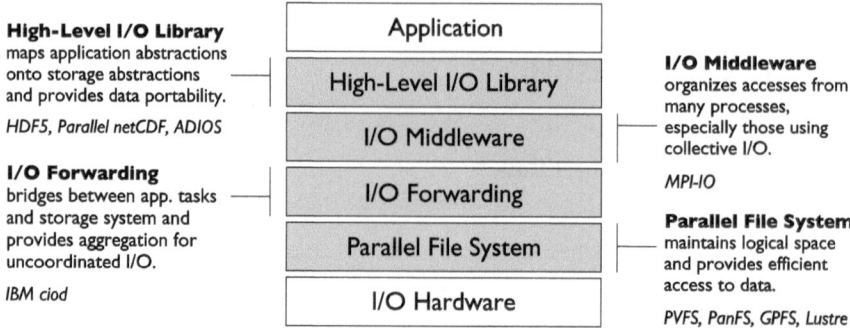

High-Level I/O Library
maps application abstractions
onto storage abstractions
and provides data portability.

HDF5, Parallel netCDF, ADIOS

I/O Forwarding
bridges between app. tasks
and storage system and
provides aggregation for
uncoordinated I/O.

IBM ciod

I/O Middleware
organizes accesses from
many processes,
especially those using
collective I/O.

MPI-IO

Parallel File System
maintains logical space
and provides efficient
access to data.

PVFS, PanFS, GPFS, Lustre

Fig. 2.1 I/O software stack. The higher levels are tailored for applications while the lower levels combine and maintain multiple channels to storage for fast parallel access

this I/O software stack, an application scientist can access I/O resources through interfaces better suited to a particular domain, while lower levels of the I/O software stack carry out optimizations and deal with storage-specific matters behind the scenes. Although application scientists do not directly interact with these lower, more device-oriented layers, a deeper understanding of these libraries can be helpful when the time comes to tune an application's I/O, and in fact should drive code design for I/O-heavy applications.

At the highest level, a computational model operates on structured data, usually a multidimensional grid of latitude, longitude, altitude, and quantities such as temperature or pressure, or possibly an even more elaborate structure. This application model fits the science and has very little, if any, relationship to the way storage is organized.

Every application will at some point need to carry out I/O, either to read in a data set or to write out a checkpoint or intermediate result. Ideally it would do so while maintaining the data structures used for the science model. Further, scientists need to collaborate with other groups or run on different computer systems. Data format libraries (often referred to as high-level I/O libraries) solve both these issues. We will cover these libraries in more detail in Sect. 2.4.

Isolating data format libraries from the file system, the I/O middleware (e.g., MPI-IO, the I/O layer of MPI-2, The MPI Forum 1997) component introduces several approaches for achieving performance portability. Collective I/O and MPI datatypes can help extract maximum performance from the underlying file system. Application scientists, however, often find these constructs a poor fit for their scientific models. Sect. 2.3 provides greater detail about the MPI-IO optimizations and how the libraries built on top of MPI-IO present these features to scientists.

Parallel file systems (Sect. 2.2) are currently one of the base technologies of scientific data storage. Parallel file systems aggregate multiple storage devices into a single, unified, high-performance storage space. The file systems have a linear data layout, in contrast to the structured data of scientific applications. Parallel file systems do not have the concept of collective I/O, though some might have data access routines capable of expressing fairly sophisticated data access methods. While several abstraction layers insulate the application author from any specific parallel file system, knowledge about the underlying file system can still give a clue as to which access patterns are more likely to perform well.

Having these multiple layers of abstraction between scientific applications and the storage subsystem can complicate matters, but each layer provides key features needed to achieve the highest I/O performance. Parallel file systems provide a general-purpose interface in addition to managing storage devices, allowing for a broad array of uses. Middleware libraries can target multiple file systems but still need a Application Programming Interface (API) broad enough to apply across a variety of science domains. Data format libraries can dictate programing structures and file formats tailored for their end users; and because multiple data format libraries can target a single middleware library, each application domain can have a data format library that best meets its needs.

2.2 Parallel File Systems

Many different file systems provide shared access from a collection of clients, spanning the range from centralized to completely multiplexed. One way to categorize these is by the degree to which they enable efficient concurrent access. At one end of the spectrum are parallel file systems (PFS), designed for concurrent access from hundreds or thousands of processors to a single file. At the other end of the spectrum, a distributed file system like NFS[1] is more of a connectivity solution, providing access from many clients, but not performing well when many accesses take place concurrently. Between these two are what are sometimes termed cluster file systems, such as CXFS (Shepard and Eppe 2006). These file systems provide data distribution that enables some level of concurrent access, but they are not architected to support the high degree of concurrency seen in high-performance computing (HPC) workloads.

In many ways a parallel file system acts just like any other file system. The same tools to create files, make directories, and read and write data work on both serial and parallel file systems. This intentional similarity allows legacy applications and familiar tools to seamlessly operate alongside high-performance applications.

The biggest contribution a parallel file system makes to the software stack is its aggregation of multiple network links, servers, and disks into a global namespace. This aggregation provides a convenient logical "unit" that applications can interact with. Figure 2.2 depicts how a parallel file system might distribute a large file across the servers to enable concurrent access. Clients can communicate concurrently with the servers when they read or write different portions of a file. Thus they are capable of achieving better performance than a single server could provide, because the load

---

[1] Actually NFS isn't a file system at all, but rather a file system *protocol*; but for our purposes in this section it is easiest to just think of it as a file system.

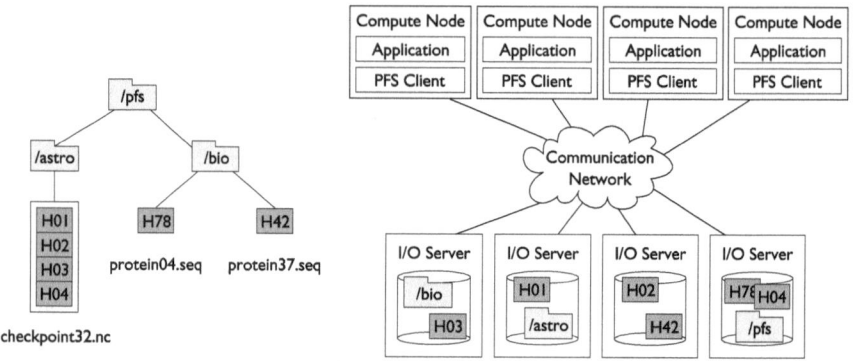

Fig. 2.2 Major components of a parallel file system: hierarchical namespace for stored files, simultaneous access by many clients, shared interconnect, and multiple server nodes managing persistent storage. Small files may be placed entirely on one server, while large files are broken up and scattered across multiple servers

is distributed across many servers, disks, and network links. The major parallel file systems today are Lustre (Braam 2003), GPFS (Schmuck and Haskin 2002), PVFS (PVFS development team 2008), and PanFS (Nagle et al. 2004).

Parallel file systems manage the data on disks. Two main designs exist: block-based and object-based. Block-based parallel file systems have their roots in older storage system designs, with GPFS being a current example of this design. Data on disk is globally allocated and managed in blocks, often many kilobytes or megabytes in size. Should an I/O request only partially fill a data block, the file system still operates on the entire block. For reads, this poses little problem: the software can merely discard any unnecessary data. Writes become trickier: the file system must carry out a *read-modify-write* sequence. First the block is read into memory, then the modifications are made, and finally the buffer is written out to disk.

In parallel applications, care must be taken to prevent *false sharing* during writes, when two or more processes want to modify different regions of the same data block at the same time. Locking mechanisms are deployed to control such problems, but at a cost of serializing what could have otherwise been a parallel data access. Small I/O requests are never good for a parallel file system, but users of block-based file systems would be well advised to align accesses to block boundaries (the other layers in the software stack often do so automatically) and to read and write in multiples of the block size.

More recently, parallel file systems have adopted an object-based storage model (e.g., PVFS, Lustre, PanFS). In this scheme, clients operate not on blocks but on higher-level abstractions called *objects*. This object abstraction imposes some structure on the data in the storage devices and gives the storage servers more intelligence about how clients are accessing data. Additionally, this approach allows storage servers to locally manage allocation of space on disks, distributing the overhead of this process. The object-based storage model, with a more sophisticated client-server

API and more flexible servers, is an important step toward addressing the problems of false sharing and read-modify-write overhead. While the underlying storage model might still be block-based, the object layers provide scope for optimizing low-level data access.

2.3 MPI-IO

In the early days of supercomputers, each vendor had its own communication library, and moving a code to a different supercomputer meant rewriting a significant portion of that code. The MPI Forum came together as a response to this byzantine landscape and proposed a single, standard message passing interface. Each supercomputer vendor implemented MPI, and as a result programmers could go back to writing scientific codes instead of worrying about machine-specific communication interfaces.

The MPI Forum reconvened a few years later to round out the MPI standard with additional features (The MPI Forum 1997), including a section on I/O, often called MPI-IO. Just as MPI provides an abstraction for communication networks, MPI-IO provides a common interface to both parallel and legacy file systems. Today, nearly every MPI library also contains an implementation of MPI-IO.

The MPI-IO API feels a lot like message passing, but for files. Sending a message becomes a write to a file (likewise for receiving/reading). MPI-IO brings two important concepts from MPI to the I/O software stack: noncontiguous and collective access. These features allow for a wide range of optimizations in the MPI-IO library, such as two-phase I/O (Thakur et al. 1999) or data shipping (Prost et al. 2001), that can significantly improve access rates in scientific codes. Depending on the level of file system support, a sophisticated MPI-IO implementation can turn what would normally be a poorly performing workload into something close to the ideal (but uncommon in scientific applications) large-block sequential I/O (Ching et al. 2002, 2003a, b).

File system APIs typically offer only a linear view of data: all data fits into one dimension, from byte 0 at the beginning of the file up to the last byte of the file. Scientific applications, however, have more sophisticated data structures. Even something as simple as reading a column out of a row-major array turns a logically contiguous access into one that is scattered across the file. This noncontiguous I/O access is fairly common in scientific applications.

In the same way that programmers use MPI datatypes to describe communication patterns, programmers can also use MPI datatypes in MPI-IO to describe noncontiguous data in memory and define a file view with these datatypes to describe the layout of data in the file.

Even if the file system does not support such a rich data access API, the MPI library can still optimize these noncontiguous accesses. With data sieving (Thakur et al. 1999), a single, large I/O operation replaces several smaller ones. Like a sieve, unneeded data is discarded. This optimization works well if the "density" of the request is high—if there is not much wasted data. In the write case, just like a block-oriented file system, the programmer must be vigilant to prevent false sharing during the read-modify-write phases, ensuring no concurrent operations within a block boundary, if high performance is the goal.

The MPI-IO layer introduces another important optimization: collective I/O. A parallel program often has distinct phases where all processes are involved in an I/O call, such as when writing out a checkpoint file or reading in a data set. If all processes access the file system independently, this storm of uncoordinated requests can be hard to service efficiently. If all processes instead carry out I/O in concert, the MPI-IO collective I/O routines can aggressively optimize the access pattern.

Collective I/O yields four key benefits. First, optimizations such as data shipping and two-phase I/O rearrange the access pattern to be more friendly to the underlying file system. Second, if processes have overlapping I/O requests, the library can eliminate duplicate I/O work. Third, by coalescing multiple regions, the density of the I/O request increases, making the two-phase I/O optimization more efficient. Fourth, and perhaps most important for scaling to the largest supercomputers, the I/O request can also be "aggregated" down to a number of nodes more suited to the underlying file system. These optimizations make some form of collective I/O almost compulsory for any application requiring high I/O performance at large scale (Yu et al. 2006). Examples of some of these techniques are to be found below in Chap. 3.

In order to tune behavior of these optimizations, MPI-IO provides a way for applications to pass parameters down to the library. These key-value string pairs, called hints, provide a way for developers to inform the MPI-IO library about application-specific workloads. For example, one could forcibly disable the data sieving optimization if the access pattern was going to be very sparse, or tune the intermediate buffer to be exactly the same size as a record of interest. Gropp et al. (1999) provides further information about MPI-IO hints.

With data types, file views, and hint parameters, one might consider the MPI-IO interface overwhelming for a scientific programmer. MPI-IO makes key optimizations *possible*, but it does require understanding a fairly complex API. Fortunately, scientists have access to several libraries built on top of MPI-IO to bridge this "usability gap" by hiding the details of using MPI-IO and presenting an interface more suited for applications.

2.4 Data Format Libraries

While applications have a computational model that operates on physical elements, PFS and MPI-IO interfaces work in terms of bytes in files. Converting between these two models can be a complex and tedious process. Furthermore, collaboration with other scientists means data exchange and can be facilitated by the use of a common data representation. Data format libraries (often called high-level I/O libraries because they are built on top of MPI-IO) address both of these issues.

Data format libraries bridge the gap between scientific applications and MPI-IO or lower levels of the I/O software. These libraries present an interface based on multidimensional arrays of typed variables. For example, when modeling atmospheric phenomena, a multidimensional array holds values for latitude, longitude, altitude, and some attribute like temperature or barometric pressure. Multidimensional arrays are intended for structured data, where contiguous array indices indicate nearest-neighbour relationships; these assumptions don't map perfectly to some approaches such as unstructured grids or adaptive meshes, but they do match well to common models and provide a good building block for more sophisticated abstractions.

Metadata—information about data such as timestamps, provenance, and workflow processes—can be as important as the actual data. High level libraries provide methods to annotate variables and entire data sets with attributes. With these annotations in place, collaborators (both now and in the future) have more context in which to understand the data.

Parallel NetCDF (Li et al. 2003) and HDF5 (The HDF Group 2008) are the commonly used higher-level parallel I/O libraries today. Both share many of the same features. Consider the NetCDF example in Fig. 2.3. The dataset contains multidimensional, typed variables. Both the variables and the dataset itself can have additional attributes. The resulting file, thanks to a well-defined file format, contains all the information needed to read the data on any platform. Recently, members of the serial NetCDF project introduced a third parallel I/O library. They implemented the serial NetCDF API using the HDF5 library. The new NetCDF-4 library (Rew et al. 2006) can use several HDF5 features, including parallel I/O. NetCDF, HDF and other data formats are discussed in greater detail in Chap. 5.

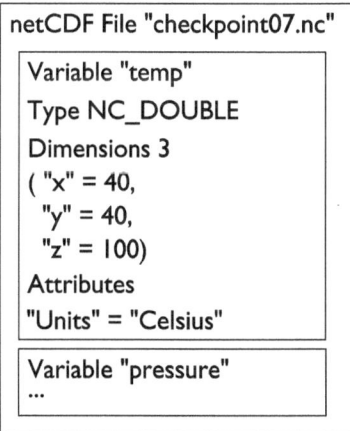
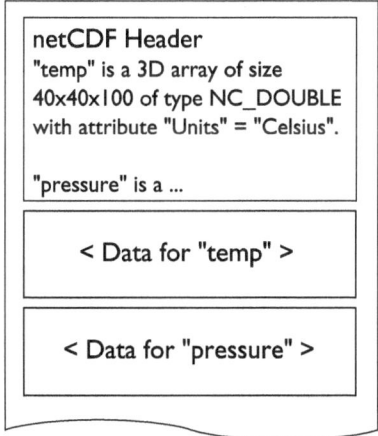

Fig. 2.3 NetCDF: one example of a data format library. The user treats data stored in NetCDF as a series of multidimensional arrays of typed data. These arrays have names and other metadata. The library manages the actual layout on disk, allocating space for a header as well as each variable. Note that both Parallel NetCDF and NetCDF use this same file format

Even though a high-level library requires another set of functions between an application and the underlying MPI-IO library, this extra layer introduces negligible overhead when used appropriately. In some cases, in fact, the abstraction allows for more aggressive optimizations, because the library API provides a richer description of the user's desires than do the lower-level APIs.

For example, the Parallel NetCDF library knows how to cache the file header information, but the MPI-IO layer is unable to distinguish the file header from any other region of the file. The HDF5 library uses a sophisticated allocation routine to optimize sparse matrix operations.

Section 2.3 covered some of the ways MPI-IO makes use of MPI datatypes to describe I/O access patterns. These datatypes map well to the multidimensional accesses used in data format libraries. The challenge of using MPI-IO directly lies in the relative complexity of describing these memory and file regions. Data format libraries manage the use of MPI datatypes on behalf of the application, allowing for a host of optimizations at the MPI-IO level without effort from the user.

As discussed earlier, scientific applications possess distinct periods of computation and I/O. Data format libraries make it easy to carry out the I/O in these phases collectively. When applications use the collective I/O routines in these higher level libraries, the MPI-IO implementation can do much to manipulate the I/O requests into the most appropriate format for the underlying storage system. As with and collective I/O, single-file I/O means more opportunities for optimization and a better chance at achieving higher I/O rates. For example, noncontiguous requests from different clients to adjacent regions can be coalesced.

Space limitations enforce only a brief treatment of I/O libraries. Chapter 5 will discuss more about the file formats in use in environmental models. The bibliographic references for these libraries will also lead to more in-depth materials.

2.5 Applying I/O Lessons to Applications

The biggest payoffs with respect to parallel I/O performance come from a simple rule: Perform I/O with the fewest calls, and pass as much information as possible about the intended results. Usually this means describing I/O as a single collective I/O call to or from one file for all processes. This rule gives the I/O software stack the most opportunity to optimize the I/O access pattern. Naturally, this single piece of advice won't work for all situations and might not yield all the performance gains one could expect, but it will go a long way toward that end.

Nothing harms parallel I/O performance more than small I/O requests. No matter what level of the I/O software stack, an application should describe all the I/O in as few calls as possible. MPI datatypes or the datatype features of the data format libraries (e.g., HDF hyperslabs) can select all data with a single function call instead of making many individual calls for one element at a time. Applications may benefit by setting aside some memory to combine accesses, though the lower layers of the software stack might already offer a tuned and debugged version of this optimization.

Awareness of the underlying file system might guide developers toward an ideal I/O strategy. Some file systems make extensive use of locks to enforce Portable Operating System Interface (POSIX) consistency semantics. Unfortunately, these lock-based file systems impose serialization of access and undermine most of the content in this section. The locking mechanisms are conservatively designed for safety; application scientists should investigate ways to disable locks for their site, at least for applications where it is semantically possible.

All the levels of the software stack have ways to give "hints" about what an application wants to do. These hints are sometimes simple, such as opening a file as either read-only or write-only but not both. Other hints are more complex, such as using MPI-IO Info parameters to adjust internal algorithms.

2.6 Parallel I/O Today and Tomorrow

From parallel file systems managing physical disks, to high-level abstractions, the I/O software stack includes a great deal of software, all of which contributes to high performance and convenient parallel I/O. The more contextual information each layer has, the more optimizations that layer can perform on behalf of the application. In fact, there is more to the I/O stack than could be covered in this space, and still more under development.

One topic that we did not discuss is asynchronous I/O. Asynchronous I/O interfaces provide a way for an application programmer to initiate I/O operations and then test for their completion later. Ideally the underlying I/O software would perform data transfer behind the scenes, allowing the application to go back to computation while I/O occurs. Unfortunately, in practice this is not the case. Most I/O software implements asynchronous I/O with blocking operations – either all I/O is performed on the first call, or all I/O is performed when the application tests for completion. Further, if implementations were to perform I/O behind the scenes, the I/O communication could interfere with communication performed as part of the computation, negatively impacting performance.

This said, asynchronous interfaces do provide another opportunity for libraries to combine I/O operations (e.g., as in PNetCDF), and the trend toward increasing core counts and more feature-rich networks in HPC systems may provide additional opportunities for more effective asynchronous I/O systems in the future.

The careful reader will notice that one part of Fig. 2.1 – I/O forwarding – was not discussed. I/O forwarding helps large I/O subsystems scale by hiding some number of clients behind a forwarding layer and presenting many fewer clients to the file system. This software exists only on today's very large supercomputer systems, where it is used to reduce the apparent number of clients by a factor of 25 to 250.

I/O forwarding and other techniques are expected play a larger role as HPC systems move toward even greater levels of parallelism. Buffered aggregation methods (Ma et al. 2003) have also shown promising results, and are being incorporated into libraries like ADIOS. Current research also focuses on abstractions that are even more domain specific. The unstructured grid and adaptive mesh models do not map especially well to the multidimensional array models used in current high-level I/O libraries, as indicated earlier, and solutions are being investigated. Some emerging trends are discussed below in other sections, including especially Chaps. 3 and 7.

References

Braam PJ (2003) The lustre storage architecture. Technical Report, Cluster File Systems, Inc., http://lustre.org/docs/lustre.pdf

Ching A, Choudhary A, Coloma K, Liao W, Ross R, Gropp W (2003a) Noncontiguous i/o accesses through MPI-IO. In: Proceedings of the third IEEE/ACM international symposium on cluster computing and the grid (CCGrid2003)

Ching A, Choudhary A, Liao W, Ross R, Gropp W (2002) Noncontiguous i/o through pvfs. In: Proceedings of the 2002 IEEE international conference on cluster computing

Ching A, Choudhary A, Liao W, Ross R, Gropp W (2003a) Efficient structured data access in parallel file systems. In: Proceedings of cluster 2003, Hong Kong

Gropp W, Lusk E, Thakur R (1999) Using MPI-2: Advanced features of the message-passing interface. MIT Press, Cambridge http://mitpress.mit.edu/book-home.tcl?isbn=0262571331

Li J, Keng Liao W, Choudhary A, Ross R, Thakur R, Gropp W, Latham R, Siegel A, Gallagher B, Zingale M (2003) Parallel netCDF: A high-performance scientific I/O interface. In: Proceedings of SC2003: high performance networking and computing, IEEE Computer Society Press, Phoenix, AZ http://www.sc-conference.org/sc2003/paperpdfs/pap258.pdf

Ma X, Winslett M, Lee J, Yu S (2003) Improving MPI-IO output performance with active buffering plus threads. In: Proceedings of the 2003 international parallel and distributed processing symposium, IEEE, pp 10

Nagle D, Serenyi D, Matthews A (2004) The panasas activescale storage cluster: Delivering scalable high bandwidth storage. In: SC '04: Proceedings of the 2004 ACM/IEEE conference on super-computing, IEEE Computer Society, Washington, DC, USA, p 53, http://dx.doi.org/10.1109/SC.2004.57

Prost JP, Treumann R, Hedges R, Jia B, Koniges A (2001) MPI-IO/GPFS, an optimized implementation of MPI-IO on top of GPFS. In: Proceedings of SC2001

PVFS development team (2008) The PVFS parallel file system. http://www.pvfs.org/

Rew RK, Hartnett EJ, Caron J (2006) NetCDF-4: Software implementing an enhanced data model for the geosciences. In: 22nd international conference on interactive information processing systems for meteorology, oceanography and hydrology, AMS

Schmuck F, Haskin R (2002) GPFS: A shared-disk file system for large computing clusters. In: First USENIX conference on File and Storage Technologies (FAST'02), Monterey, CA

Shepard L, Eppe E (2006) SGI infinite storage shared filesystem CXFS: A high-performance, Multi-OS filesystem from SGI

Thakur R, Gropp W, Lusk E (1999) Data sieving and collective I/O in ROMIO. In: Proceedings of the seventh symposium on the frontiers of massively parallel computation, IEEE Computer Society Press, pp 182–189, http://www.mcs.anl.gov/ thakur/papers/romio-coll.ps

The HDF Group (2008) HDF5. http://www.hdfgroup.org

The MPI Forum (1997) MPI-2: extensions to the message-passing interface. The MPI Forum, http://www.mpi-forum.org/docs/docs.html

Yu H, Sahoo RK, Howson C, Almasi G, Castanos JG, Gupta M, Moreira JE, Parker JJ, Engelsiepen TE, Ross R, Thakur R, Latham R, Gropp WD (2006) High performance file I/O for the blue-gene/l supercomputer. In: Proceedings of the 12th international symposium on high-performance computer architecture (HPCA-12). http://www.mcs.anl.gov/ thakur/papers/bgl-io.pdf

Chapter 3
ESM I/O Layers

V. Balaji

3.1 ESM I/O Layers: Design Considerations

While Earth System models have been among the very largest consumers of high-performance computing (HPC) cycles, they very rarely figure on lists of computing *tours de force*, such as the Gordon Bell prize. Among the reasons are that ESMs, especially climate models, tend to be among the most I/O intensive applications in HPC.

There are several challenges associated with I/O from ESMs.

- ESMs, by their very nature, are tools to understand the long-term and large-scale effects of phenomena spanning downward to fine spatial and short temporal scales. A span of 8 orders of magnitude is not uncommon: ESMs are often integrated for $\mathcal{O}(10^6) - \mathcal{O}(10^8)$ timesteps and contain $\mathcal{O}(10^6) - \mathcal{O}(10^8)$ grid points. The reason is that there is basic physics at small space and time scales that has cumulative effects on global climate equilibria and secular climate change. For instance, in a pioneering high-resolution ocean simulation to understand the effect of small-scale ocean eddies on ocean climate, Hallberg and Gnanadesikan (2006) found 5-day 3D snapshots of the global eddy field of 10^8 grid points saved over decades-long runs were necessary to compute eddy statistics necessary to understand the associated climate processes. Similarly, studies of cloud forcing (e.g Williams et al. 2006) request 3D snapshots of the cloud field every hour over periods of decades.
- The current practice in Earth system modeling is to run large coordinated international modeling campaigns, where similarly configured experiments are run by modeling centres around the world, such as the Intergovernmental Panel on Climate Change (IPCC) earlier alluded to. This methodology requires that the centres adhere to *metadata standards and conventions*. I/O layers that provide means to

V. Balaji
Princeton University, Princeton, USA
e-mail: balaji@princeton.edu

V. Balaji et al., *Earth System Modelling – Volume 4*, SpringerBriefs in Earth System Sciences, DOI: 10.1007/978-3-642-36464-8_3, © The Author(s) 2013

easy production of standard-compliant data are an essential component of ESM I/O.

- *Postprocessing* of data generated from a running model is a ubiquitous element in the I/O workflow. An ESM simulation may consist of a sequence of multiple runs, and a subsequent analysis may require aggregating data from these multiple runs.

These considerations have led to the development of ESM I/O and postprocessing layers, outlined in this section. The basic machinery of parallel I/O was described in chap. 2. In Sect. 3.2 we describe a typical ESM I/O layer called mpp_io providing simple data structures for adding metadata standards and conventions to parallel I/O libraries. This layer also provides access to multiple data formats, described below in chap. 5. In Sect. 3.3 we describe toolsets associated with the NetCDF data format most commonly used in Earth system modeling for the postprocessing of data. These tools provide means to manipulate model output to enable participation in multiple modeling campaigns.

3.2 An ESM Parallel I/O Class Library

The basic machinery of parallel I/O was described in Chap. 2. In Sect. 2.4 a description of high-level libraries for parallel I/O in specific ESM community standard data formats such as NetCDF were also described. While these libraries are an essential part of the I/O software stack, they do not specify how physical information associated with the data, for example, what grid the quantity being written is described upon, or what physical quantity the saved data represents, or what its units might be – these are data attributes upon which those libraries are silent.

Parallel I/O has been an important consideration in the development of *frameworks*, described in Volume 1 of this series, some of which have included abstractions for simplified interfaces that include metadata to describe the physical information. Typical abstractions used to capture this information are two classes or objects: *grids* and *fields*.

- Grids capture the physical locations and structure of grid points upon which data are stored. This may include geo-referencing information (latitudes and longitudes) to relate data points to the planetary surface: however, the grid class has to be general enough to support cartesian grids and other arbitrary surfaces. Since this is associated with a parallel application, the grid class is usually generalized to a *distributed grid* class, that in addition to the gid information, also contains information about the domain decomposition and layout of the grid across processors.
- Fields are a container class for physical quantities. In addition to the actual data array holding the values of a physical variable, the class holds metadata describing the variable, giving it a name (e.g "Temperature"), sometimes a name drawn from a standard name table ("air_temperature") such as the one associated with the

Climate and Forecast (CF) Conventions, as well as other attributes identifying its units ("kelvin"), the valid range of values, etc.

Many current I/O packages used in ESMs today use these classes to enable an I/O class library. Examples include `mpp_io`, used by the Flexible Modeling System (FMS) and the Partnership for Research Infrastructure in Earth System Modeling (PRISM) frameworks (see Volume 2 in this series); CF-IO, developed at the National Aeronautics and Space Administration (NASA) Global Modeling and Assimilation Office (GMAO) and following Earth System Modeling Framework (ESMF) bindings; WRF-IO, used by the Weather Research and Forecasting model (WRF) that is used widely in the numerical weather prediction (NWP) community, and the Adaptable IO System (ADIOS, Lofstead et al. 2008). We briefly describe some features and call sequences in `mpp_io`, which may be considered somewhat typical.

3.2.1 *mpp_io*

`mpp_io` supports four I/O *modes*. These are described here in the output mode, but the converse process works for input.

- *Single-threaded I/O*: from a parallel code, this implies that the global array is gathered onto a single processor for output. There is one I/O thread per model component, so there can still be multiple I/O streams from an application.
- *Multi-threaded I/O*: multiple processors write into a single file. This requires an underlying parallel I/O library as outlined in Chap. 2.
- *Distributed I/O*: each processor writes to its own file, which must then be later knitted into a single file. The `mpp_io` package writes extra metadata to enable the later combination step. The package comes with a tool, `mppnccombine`, that effects this step for NetCDF files.
- *Quilted I/O*: borrowing a term from WRF-IO, this method combines aspects of single-threaded and distributed I/O. Rather than gather the data onto a single processor, we aggregate the data onto a few (a "few" being $\mathcal{O}(10)$–$\mathcal{O}(100)$ fewer than the parent component) processors, which then perform distributed I/O.

The four modes are necessary to accommodate different data formats, including those with no underlying parallel I/O library, as well as to accommodate filesystems with varying capabilities.

Besides the modes described, the package supports multiple data *formats*: ASCII formatted output, the binary format native to the hardware system, the Institute of Electrical and Electronics Engineers (IEEE) standard floating-point format at 32- and 64-bit, NetCDF-3 and -4.

The abstractions of distributed grids and fields; the definition of the different parallel I/O modes; the different output formats; all together allow the definition of an I/O class library. We present here the `mpp_io` package as an example of what such a class library might look like:

```
type(domain2D) :: domain
type(axistype) :: x, y, z, t
type(fieldtype) :: field
integer :: npes, unit
character*(*) :: file
real, allocatable :: f(:,:,:)
call mpp_define_domains( (/1,ni,1,nj/), domain )
call mpp_open( unit, file, format=MPP_IEEE32, threading=MPP_SINGLE )
call mpp_write_meta( unit, x, 'X', 'km', ... )
    ...
call mpp_write_meta( unit, field, (/x,y,z,t/), 'Temperature', ... )
    ...
call mpp_write( unit, field, domain(pe), f, tstamp )
call mpp_close(unit)
```
$$\tag{1}$$

Code Block 1 is interpreted as follows:

- We have defined 3 container classes to hold the descriptive information about domain decompositions, physical grid locations, and physical variables: domain2D, axistype, and fieldtype.
- The global domain consisting of ni × nj points, is decomposed across npes processors, with the resulting domain decomposition stored in domain, an instance of the domain2D class.
- The call mpp_open opens a file for writing IEEE 32 bit floating point format data using the single-threaded output mode, returning a filehandle in unit. Note that if we wished to write the same data in a different format, or using a different I/O mode, only one argument to this subroutine would change, and the rest of the code remain the same.
- The physical grid is described by variables of class axistype. The data and descriptive metadata for each axis is written in a series of mpp_write_meta calls.
- The physical variable field(x,y,z,t) is declared in the next call, along with its descriptive metadata.
- After all the metadata has been written, the actual data is written. The data is held in a simple *local* 3D array f(i,j,k) holding the values at time index tstamp. The call mpp_write now has all the information it needs for a parallel write according to whichever mode was chosen in the mpp_open call.

This deceptively simple parallel I/O library, consisting of three derived types, and about 4 basic calls, represents an example of an ESM parallel I/O layer to be used by running models. It is used in the FMS and PRISM frameworks described in Volume 1. Its performance on distributed I/O has been shown to be limited only by hardware limits: for large arrays in the distributed write mode, it is only limited by number of concurrent I/O channels that can be supported by the underlying filesystem. mpp_io poses no limits on scalability of I/O-intensive codes up to $\mathcal{O}(1000)$ processors; however, as we move into regimes where the number of processors far exceeds the number of I/O channels, it begins to show its limits.

Other elements of the API are geared toward performance analysis. It is to be noted that I/O performance analysis is a ubiquitous issue: while it may be possible to "benchmark" basic I/O kernels on a quiescent system, real I/O performance varies tremendously across systems and even in time, depending on total system load. It is important to have the capability to monitor I/O performance in real time on a running application. The ability to monitor I/O volumes, latency and speed from any code section is also part of mpp_io.

mpp_io makes it very easy to produce output datasets in several formats, and follow community naming conventions such as CF (Lawrence et al. 2005).

New projects underway in mpp_io include the development of an *asynchronous* I/O layer. Asynchronous I/O permits a set of processors to perform I/O concurrently with computations elsewhere. This is implemented in mpp_io by designation an I/O domain within the domain type. The I/O processors skip the computational portions of the code: however data designated for I/O is transferred to the I/O nodes by quilting. The quilting can take the form of non-blocking messages, or take advantage of shared-memory if available. The I/O from these nodes proceeds concurrently with the computation. This is schematically shown in Fig. 3.1.

3.2.2 Other ESM Parallel I/O Libraries

Other I/O packages in the community approach include WRF-IO (Michalakes et al. 2004) and CF-IO.[1] The latter is a set of bindings very similar to mpp_io and using ESMF derived types to describe fields, grids, and decompositions.

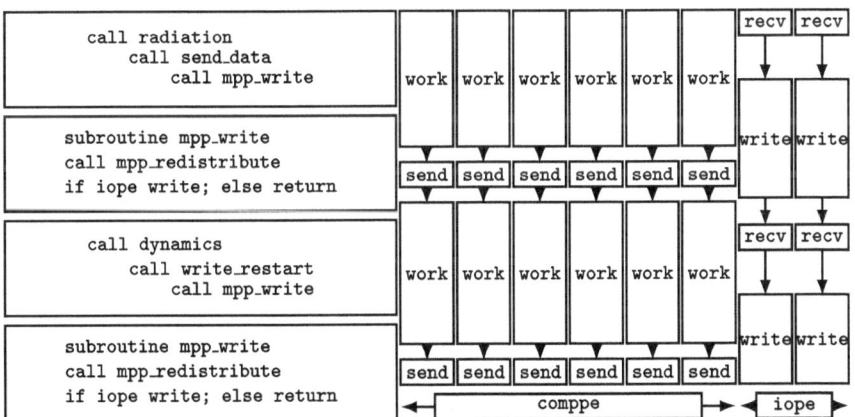

Fig. 3.1 Code flow with asynchronous I/O domains. This is shown schematically for 6 computational PEs and 2 I/O PEs. The redistribute calls (shown as send/recv pairs) are non-blocking and transfer data from computational PEs to I/O PEs

---

[1] http://www.maplcode.org/maplwiki/index.php?title=MAPL_I/O.

Finally, the ADIOS package (Lofstead et al. 2008) is a promising new development in this field that might overcome the scalability limits of `mpp_io` and similar packages. It is tightly coupled to the parallel I/O layers described in Chap. 2. ADIOS too has an extremely simple API and is configured through external Extensible Markup Language (XML) based configuration files, which makes it easy to integrate into workflow tools, such as those described in Volume 5 of this series. These libraries now include extensive support for formats relevant to the ESM community, such as NetCDF. Work in this area is active and underway.

3.3 ESM Post-Processing Tools and Libraries

Post-processing consists of the computations that are done after the model is run. For example, one might wish to regrid output from the model's native grid onto some standard spatial grid; compute statistics; derive auxiliary variables from saved variables.

The output from an ESM simulation might be considered a 6-dimensional variable $f(x, y, z, t, v, n)$, where the simulation consists of an *ensemble* of runs, with the ensemble number here denoted by n; where many variables are saved from each simulation, here denoted by v; and for each variable, the data consists of 3 space and one time dimension. Each physical file output from a run consists of some hyperslab within this 6-dimensional space.

The I/O layers described in Sect. 3.2 are used to write data out from a running Earth System model. Consider that an ESM run might last centuries or millennia of model time, and days or months of wallclock time; however, a single run or job is often a small fraction of that. Output must be synchronized to persistent storage in filesystem units necessarily limited to contain no more data than that produced from a single run.

However, the kinds of analysis described in Chap. 6 might require a complete time series of a model run. The construction of a complete time series requires an additional step of *aggregation*: the creation of a "virtual file" that provides a way to access a complete time-series of data stored across multiple physical files.

One of the major complications associated with post-processing is the aggregation problem. While protocols such as OPeNDAP (Cornillon et al. 2003) offer on-the-fly aggregation, the performance of such protocols falls well short of what is needed.

The analysis of ESM model output might require aggregation along one or more of these axes. For instance, one might aggregate along time in order to compute a climatological mean of some field; one might aggregate multiple variables in order to compute some derived field; one might aggregate across n to compute ensemble statistics. In fact, distributed output from a parallel run might be disaggregated in space and require aggregation.

Post-processing I/O layers may be composed of several steps:

- *Aggregation.* This consists of the creation of the hyperslabs necessary for subsequent computations, such as the computation of time averages. Aggregations typically perform no computations, but reorganize data stored across multiple files.
  The NCO utilities (http://nco.sourceforge.net/) provide several utilities for aggregation and disaggregation.
- *Regridding and averaging.* These steps consist of transforming the data onto standard model grids, as well as constructing time averages or ensemble averages. Horizontal regridding is performed by several packages, the most common among which is the Spherical Coordinate Remapping and Interpolation Package (SCRIP Jones 1999). Other packages have implemented SCRIP algorithms, e.g. the Climate Data Operator (CDO.[2]) The `fregrid` package implements SCRIP and other algorithms using a standard NetCDF description of grids known as Gridspec (Balaji et al. 2007). CDO also provides many other utilities for analyzing climate model data.
- Finally, there are packages that organize climate model data for international modeling campaigns. The key problem addressed here is to enforce adherence to common conventions and standards from multiple models from diverse institutions. The most prominent example of this is the Climate Model Output Rewriter (CMOR) package[3] used to organize data for the Coupled Model Intercomparison Project. The success of CMOR in IPCC AR4 has to led to its adaptation for various other international modeling campaigns.

3.4 Discussion

We have described here the various layers of I/O processing associated with Earth system modeling. One layer is directly associated with a running model: ingesting data and writing data from an ESM running on parallel hardware. While underlying data transfer protocols can be complex, ESM models rely upon abstract I/O layers that provide simple means to read and write "fields" on distributed arrays on parallel hardware. These layers also provide the mechanisms to add descriptive metadata to fields. A second layer of I/O is performed after the model has been run. These tend to be more data-intensive and involve aggregation, regridding, and statistical operations. Many standard packages of operators are now available for post-processing, e.g., NCO (Zender 2008) and CDO (Schulzweida et al. 2006).

Finally, there are I/O layers for organizing the output from diverse models for comparative analysis. It is now becoming common practice to impose common organization and convention upon such massive distributed ESM data archives, to

---

[2] http://www.mpimet.mpg.de/cdo.

[3] http://www2-pcmdi.llnl.gov/cmor.

be described in Volume 6 of this series. Packages for ensuring and imposing confor-mance is also now becoming commonplace.

The I/O problem in the ESM context must thus be seen as an end-to-end problem composed of model input and output, post-processing, and distribution into wide-area data archives. An exciting and challenging area of research and study of the integrated I/O problem is underway.

References

Balaji V, Adcroft A, Liang Z (2007) Gridspec: a standard for the description of grids used in earth system models

Cornillon P, Gallagher J, Sgouros T (2003) OPeNDAP: accessing data in a distributed, heteroge-neous environment. Data Sci J 2:164–174

Hallberg R, Gnanadesikan A (2006) The role of eddies in determining the structure and response of the wind-driven southern hemisphere overturning: results from the modeling eddies in the southern ocean (MESO) project. J Phys Oceanogr 36(12):2232–2252

Jones PW (1999b) First- and second-order conservative remapping schemes for grids in spherical coordinates. Mon Wea Rev 127(9):2204–2210

Lawrence B, Drach R, Eaton B, Gregory J, Hankin S, Lowry R, Rew R, Taylor K (2005) Maintaining and advancing the CF standard for earth system science community data

Lofstead J, Klasky S, Schwan K, Podhorszki N, Jin C (2008) Flexible IO and integration for scientific codes through the adaptable IO system (ADIOS). In: Proceedings of the 6th international workshop on challenges of large applications in distributed environments, ACM New York, pp 15–24

Michalakes J, Dudhia J, Gill D, Henderson T, Klemp J, Skamarock W, Wang W (2004) The weather research and forecast model: software architecture and performance. In: Proceeding of the eleventh ECMWF workshop on the use of high performance computing in meteorology, pp 25–29

Schulzweida U, Kornblueh L, Quast R (2006) CDO users guide. Climate data operators, version 6 1(6), http://code.zmaw.de/projects/cdo

Williams K, Ringer M, Senior C, Webb M, McAvaney B, Andronova N, Bony S, Dufresne J, Emori S, Gudgel R et al (2006) Evaluation of a component of the cloud response to climate change in an intercomparison of climate models. Clim Dynam 26(2):145–165

Zender C (2008) Analysis of self-describing gridded geoscience data with NetCDF operators (NCO). Environ Modell Softw 23(10):1338–1342

Chapter 4
Data Storage

Bernie Siebers and V. Balaji

4.1 Hierarchical Storage Management: An Introduction

Modeling entails storage of large quantities of data for varying lengths of time. It is often difficult to know when a new analytical tool or scientific insight may necessitate reanalyzing model output in a new way. It is infeasible to save every variable at every timestep, so modelers must frequently weight the cost of saving a bit of data which might never be used against the cost of rerunning portions of the simulation if an unsaved bit is needed.

A key to enabling the large-scale storage needed is minimizing the cost of saved data and maximizing the convenience of retrieving it. As convenience usually comes at a price, this calls for striking an appropriate balance. One tool which helps manage that balance is an HSM, or Hierarchical Storage Management system. Some example of widely used HSMs in High Performance Computing are IBM's High Performance Storage System (HPSS, Watson and Coyne 1995), SGI's CXFS and Data Migration Facility (DMF) (Shepard and Eppe 2003) and Oracle's Storage- und Archive-Manager Quick File System (SAM-QFS) (Read 2010).

An HSM manages files apparently resident in one or more filesystems, which are accessible directly or via ftp, by moving them "under the covers" to and from other storage devices and types, for example tape. These other storage media are organized into levels. The filesystem of apparent residence is the top level, and usually has high bandwidth, low latency and/or high duty cycle, compared to the other levels. The next level may still be on disk, but with lesser performance characteristics and lower cost. The lowest levels are typically on tape, perhaps of more than one type, robotically

B.Siebers
Geophysical Fluid Dynamics Laboratory, Princeton, USA
e-mail: Bernie.Siebers@noaa.gov

V. Balaji
Princeton University, Princeton, USA
e-mail: balaji@princeton.edu

V. Balaji et al., *Earth System Modelling – Volume 4*, SpringerBriefs in Earth System Sciences, DOI: 10.1007/978-3-642-36464-8_4, © The Author(s) 2013

managed or on shelves. An individual file may reside simultaneously at multiple levels, or be copied more than once to the same level. Configuring an HSM to keep two copies of tape data is common, but not universal, to reduce the chance of lost data.

HSM software moves the files from one level to another automatically, based on file characteristics such as access time, size and ownership, or manually, pursuant to user or administrator commands. The objective is to place files which have high priority or will likely be needed soon on high performance storage; to store files which may never be accessed again but must be saved for reference purposes as cheaply as possible, and perhaps to store files between those two levels of interest on storage media with intermediate cost and performance characteristics. An HSM thus behaves much like a cache manager, but may use a more complex aging or weighting scheme than the typical cache manager, and often permits more manual intervention.

An HSM is designed to allow a user to access a managed file without knowing its level of residence, but sometimes it is good to know. Suppose you are running a long model, accessing files in a natural sequence, letting the HSM recall files as needed, and those files happen to reside on tapes stored on racks. It could take many minutes or even hours to retrieve those files, delaying your job and potentially idling processors.

In another example of ignorance not being bliss, tape files in an HSM look superficially like disk files, so it is easy to create and delete them as if they really were on disk. However tape is fundamentally a serial medium. Deleting files resident on tape doesn't actually free up tape space. An HSM handles this by finding tapes with relatively little active data remaining. It moves all remaining data on such tapes to other volumes, and the emptied tape becomes available for reuse. While this permits more efficient space utilization of the tapes, it takes resources which would otherwise be available for current tape activity, and does so in a way which is nearly invisible to the user.

4.2 An Abstraction for HSMs: A Three-Level Storage Model

A typical user interface to hierarchical storage uses a two-level storage model: involving one set of calls to retrieve a data resource from "deep storage", and then other commands to manipulate disk-resident data. For instance, on a system with HPSS mass storage, one might issue a `hsi put` to retrieve an object from deep storage and place it on a Lustre or similar parallel filesystem. On a DMF system, the file appears disk-resident, but extended `stat` attributes identify the file as being "migrated", and then a `dmget` is issued.

As remarked earlier, linear media are markedly inefficient for non-linear use:

- random access patterns spanning non-contiguous regions of tape or multiple tapes result in multiple "tape fetch" requests;
- file deletion (creates "holes" that are later filled in background tape defragmentation activity).

With a two-level storage resource model, it has been difficult to avoid non-linear use of linear media. We propose instead a three-level model:

tmp   "fast" scratch space that is not guaranteed to exist beyond the end of an access session;

ptmp   random access storage that is not guaranteed to be backed up, and is used as a staging area or cache for data in deep storage. The "persistent temporary" storage also provides continuity of the cache between data sessions: however, data integrity between sessions may not be guaranteed, as this level of storage is usually managed with aging and scrubbing algorithms guaranteeing free space. For data that is not easily re-created, archiving is a necessary step.

archive   backed-up storage that is modeled as though it were "remote", i.e need-ing explicit fetch and store instructions.

The schematic names of tmp, ptmp and archive provide mnemonic clues to the use of that filesystem.

- The tmp resource is usually tuned for rapid multiple unpredictable access episodes, and is intended for use by the user applications.
- The ptmp resource is tuned for bandwidth, i.e large predictable data streams. It is treated as a data cache between the user application and remote storage. The ptmp layer provides a way to cache the remote data so as to avoid multiple fetches from remote storage. Applications that exploit ptmp should usually use timestamps and checksums to avoid moving data unless the target is invalid or obsolete.
- The archive resource is treated as remote storage. It may be associated with an offline medium requiring specialized fetch instructions. Usage of this resource other than for fetch or retrieve instructions is discouraged. A consequence of the three-level storage model is that it provides an elegant way to integrate a "data grid", where the archive resource is a remote entity accessed across the Wide Area Network (WAN), using remote access protocols such as the Grid File Transfer Protocol (GridFTP) or the Open-source Project for a Network Data Access Protocol (OPenDAP).
  It is usually a good idea to control the number of remote fetches, by building archive files (e.g tarfiles). The HPSS htar command is an example of software integrating this capability. Of course, the user is advised to create archive files optimized for collocation of datasets: datasets that are usually accessed by an application at the same time should be on ptmp at the same time: *temporal locality* in cache parlance. We see that the addition of the ptmp cache layer allows us to exploit a number of the lessons learned from caches in other contexts.

At GFDL, we have written simple tools[1] that allow us to use a third-level storage model in various environments: with remote storage on HPSS, DMF, and OPenDAP.

From the point of view of Earth System modeling, we can conceive of several optimizations. During "prototyping" and "development" activities, there may be no need to archive much data. Even in "production", much of what we deem to be

---

[1] e.g see http://www.gfdl.noaa.gov/~vb/fre/hsmget.html.

intermediate data may not be archived. With a sufficiently large `ptmp`, one can avoid most access to the slow `archive` resource. The size of `ptmp` becomes a critical design element in building a balanced HPC system.

4.3 Discussion: HSM Strategies

So an HSM can be a useful tool, but a bit of understanding of its internal operation can stand the user in good stead. Here are some pointers which may help you take better advantage of this tool:

- Prefetch to highest HSM level or copy to unmanaged storage any files needed for model runs
- Find out what storage levels your HSM uses, and what data resides on each level and for how long
- Find out what commands your HSM makes available for manual level placement and for displaying file residence level
- Try to avoid storing transient data at tape levels
- Try to avoid storing large numbers of very small files on the HSM; recalling them, particularly from tape, can be inefficient, and HSMs make it easy to reach very large file counts
- Compute as a point of reference the 50 % efficiency point for the offline levels, particularly tape, i.e., the size file for which half the recall time the overhead to access the file and half is the transfer itself. For example for LTO-4 tape in robotic storage the time to locate the first byte of data may be a minute and the transfer rate is mm MB/sec, so for a XX GB file, the recall time is half locate and half transfer.

References

Read T (2010) Oracle solaris cluster essentials. Pearson Education, Canada
Shepard L, Eppe E (2003) SGI infinite storage shared filesystem CXFS: a high-performance, multi-OS filesystem from SGI. White Paper 2691:08–21
Watson R, Coyne R (1995) The parallel I/O architecture of the high-performance storage system (HPSS). In: proceedings of the fourteenth IEEE symposium on mass storage systems' storage-at the forefront of information infrastructures', IEEE, pp 27–44

Chapter 5
Data Representation

Robert Drach and John Caron

5.1 Scientific Data File Formats

File formats are divided into general formats such as NetCDF, HDF4 and HDF5, which are intended to support a broad range of scientific data, and specialized formats such as Hierarchical Data Format—Earth Observing System (HDF-EOS) and GRIdded Binary (GRIB), which add special functionality tailored to a narrower range of data types. All of these store binary data for storage efficiency, are portable across machine architectures, and have application programming interfaces (APIs) to various programming languages that insulate programmers from the low-level details of the file format.

The efficiency of data access to large datasets is almost always limited by disk latency and disk bandwidth. I/O time is determined by the number of disk blocks read, the bandwidth between disk and main memory, and the pattern of disk access which determines how much time is spent repositioning the disk head, called seek time. Modern disks have become very fast and very dense, such that sequentially reading a disk can be 100 times faster than randomly reading the same amount of data, due to the time to reposition the disk head. Efficiently accessing large scientific datasets therefore requires the programmer to have some understanding of how the data is stored, and how an API call translates to disk access patterns. This unfortunately mitigates some of the advantages of an API that hides those details. However this is an overriding problem only for large datasets (GBytes or more), and for medium sized files performance is a secondary concern. However, when it is a concern, one

R. Drach
Program for Climate Model Diagnosis and Intercomparison,
Lawrence Livermore National Laboratory, Livermore, USA
e-mail: drach1@llnl.gov

J. Caron
University Corporation for Atmospheric Research, Colorado, USA
e-mail: caron@unidata.ucar.edu

V. Balaji et al., *Earth System Modelling – Volume 4*, SpringerBriefs in Earth
System Sciences, DOI: 10.1007/978-3-642-36464-8_5, © The Author(s) 2013

must be aware of how the OS/file system/disk subsystem caches disk blocks, and so optimization becomes OS dependent.

5.1.1 NetCDF

NetCDF is a very simple but general, fixed-layout file format whose basic data type is a rectangular array of primitive types, a very straightforward data model familiar to Fortran-77 programmers. To this, NetCDF adds two important features: arbitrary amounts of metadata in the form of (name, value) pairs at both the file and variable level, and shared, named dimensions which define variables' index ranges. The sharing of dimensions between variables is a simple and powerful mechanism for defining coordinate systems. The physical layout of a NetCDF file consists of a header containing the structural metadata, as well as the user defined attributes, followed by the data variables in a predictable layout that allows efficient subsetting. One of the dimensions may be designated the unlimited dimension, and variables with that dimension may grow indefinitely by appending to the file. Otherwise, the variable sizes and layouts are fixed at creation time, and while redefinition is possible, it may require rewriting of most of the file to achieve. There is no compression per-se, but a convention for storing fixed precision values and converting to floating point with scale/offset attributes is widely used. This design offers simplicity and efficiency for reading, but the fixed layout can make efficiently writing large files hard. For these reasons, NetCDF is a read-optimized format.

The NetCDF application programming interface (API) and reference C library from Unidata is a major reason for NetCDF's widespread adoption. Other language bindings are available in C++, Fortran-77, Fortran-90, Perl, Python, Ruby, MATLAB, Interactive Data Language (IDL), and others, all built on top of the C reference library. An independent implementation is available in Java, called the NetCDF-Java library, also from Unidata.

5.1.2 HDF

The Hierarchical Data Format (HDF) family of file formats has a richer data model than NetCDF at the cost of increased complexity at the storage level. HDF4 is a legacy format, and is maintained but no longer developed. In addition to rectangular arrays of primitive type, HDF4 supports compound data types similar to structs in C or rows in an relational database management system (RDBMS). This allows heterogeneous data, such as different measurements constituting a single observation, to be stored physically close on disk, which improves data access when the common pattern is to want all the data from an observation at once. HDF4 also has specialized support for variable length American Standard Code for Information Interchange (ASCII) Strings, raster images, color palettes etc. Equally important as access efficiency, data types like Structures and Strings simplify an application reading such data, and reduce the cognitive burden of programming. HDF4 supports non-contiguous storage layout options using linked lists, as well as tiled array layout called chunking, which allows the writer to optimize physical storage to the expected read access pattern. It also supports internal compression of data using zip/deflate and run-length encoding.

HDF5 is a reimplementation and extension of HDF4. The formats are incompatible, although tools exist to translate between them. HDF5 uses Btrees instead of linked lists to track internal objects. It adds features to support very large datasets, and in some ways duplicates some file system features such as groups (directories), file mounts, etc. It provides support for variable length data, mappings of integers to Strings called enumerations, and user-defined types which can be reused by more than one variable. The price of all of this functionality is a very complex file format and an API with a steep learning curve. Surprisingly, neither HDF4 or HDF5 support shared dimensions, although they do have special variables called dimension scales which provide the same functionality as shared dimensions for the common 1-dimensional case. A number of interesting packages exist that extend HDF5, including PyTables, a Python-based program that adds indexing to tables of data, similar to an RDBMS.

HDF-EOS is a special version of HDF for NASA's Earth Observing System satellite data, using both HDF4 (HDF2-EOS) and HDF5 (HDF5-EOS) as the base format. HDF-EOS is a separately developed library on top of the HDF libraries which provides specialized functionality for grid, swath, and point data, as well as metadata conventions for specialized information on these data types stored in an HDF file. The HDF-EOS designers chose to place this metadata in text blocks using a non-standard format called ODL, rather than using standard HDF metadata, preventing most users from accessing the data except through the HDF-EOS library, and adding a steep learning curve and obstacle to casual use. From a data model point of view, the main information added by HDF-EOS are shared dimensions, unambiguous data typing (grid, swath, point), and a well defined coordinate system. Despite the difficulty of access for casual users, the additional semantics added by HDF-EOS are important and necessary for correct use of the data.

NetCDF4 is an extension of the NetCDF data model and API, using HDF5 for the physical storage. It allows an application already using the NetCDF API to relink their code with the NetCDF4 library, and read NetCDF4 files with no changes to their software. By staying within the "classic" NetCDF data model, users writing NetCDF4 files can take advantage of compression and chunking data layout, which are storage details transparent to readers, and remain readable by the very large existing base of NetCDF software. Users can then migrate to new features in the NetCDF4 "enhanced" data model in an incremental way. These new features are mostly a straightforward reflection of the HDF5 data model, with a few minor restrictions. NetCDF4 files can be read through the HDF5 API, with the caveat that NetCDF4 files have specialized constructs to support shared dimensions, whose semantics HDF5 will not recognize as of this writing.

5.1.3 GRIB

GRIB (GRIdded Binary) is a specialized file format developed by the World Meteorological Organization (WMO) for the representation of gridded weather data products. A widely used standard for numerical weather forecasting, GRIB Edition 1 was formalized in 1990, and updated to Edition 2 in 2000 (subsequently renamed "General Regularly-distributed Information in Binary form"). It is a sequential, physical file format that defines the bit structure of the file, in contrast to defining an application programming interface (API) to the data. A GRIB file consists of a sequence of 2-D horizontal sections from a 3-D or 4-D variable. One of the strengths of GRIB is its support for a variety of packing and compression methods: fixed-precision bit packing, run length encoding and incremental encoding in edition 1, and sophisticated lossless wavelet-encoding (JPEG2000) in edition 2. Although it contains a description of the data grid, as well as other descriptive metadata, it is not a fully self-describing format. GRIB relies on references to external tables for the definition of the parameters (variables). In the past each major data center defined its own GRIB tables, which led to a profusion of tables and made it difficult to write and support fully general decoding software. There is hope with the introduction of GRIB2 that the tables will converge to a standard set. GRIB2 also introduces the ability to store graphics products in JPEG and PNG form. Most of the major weather data centers, such as the European Centre for Medium-Range Weather Forecasts (ECMWF) and the National Centers for Environmental Prediction (NCEP), use GRIB.

Although GRIB does not define a standard API for encoding/decoding GRIB files, there are a variety of public domain software libraries and utilities available. In addition to routines for sequential decoding, there are utilities to support indexing and direct access to one or more GRIB files. For example, the Grid Analysis and Display System (GrADS), gribmap utility creates an ASCII control file that recovers the 4-D structure of the data. The Climate Data Analysis Tools (CDAT) package also reads GRIB via GrADS control files.

5.2 Remote Data Access

Accessing data remotely has become more important as both observational and model datasets become much larger. Remote data access is roughly divided into three categories: bulk access, subsetting in index space, and subsetting in coordinate space, each requiring progressively more knowledge of the data internals by the server.

A subsetting server must understand the format of any file that it serves, create a representation of the file's contents, and allow a client to obtain the representation and make requests for data subsets. The representation is essentially the server's data model for the file.

5.2.1 Bulk Access

Bulk access refers to transferring an uninterpreted sequence of bytes, corresponding to an entire file, or part of a file. The data model is simply an array of bytes. A server that supports partial file transfer allows a client to specify a byte range, that is, an index range into the byte array.

The File Transfer Protocol (FTP) is one of the oldest and most widely used technologies for this purpose. The Hypertext Transfer Protocol (HTTP) offers much the same functionality, and is slowly supplanting FTP for simple file serving due to the ubiquity of web servers. HTTP 1.1 allows partial file serving though the Accept-Range header. GridFTP is an important extension of FTP for the Grid community, part of the Globus toolkit. It integrates with Globus security to provide authentication and encryption for the file transfer. It was designed for high-performance grid computing, and can open multiple simultaneous TCP streams, optimize TCP window sizes, resume interrupted transfers, and transfer only a portion of a file. Because it is tied into the Globus toolkit, it has not been used much outside of the Grid community.

5.2.2 Remote Subsetting in Index Space

OPeNDAP is an HTTP protocol for remote subsetting of scientific datasets. Its data model is approximately the same as the NetCDF/HDF5 model. While OPeNDAP servers provide a human-readable (HTML) interface, its main use is by remote clients that can query the server for a description of a dataset, and then request specific subsets of the data. A client uses a compact and intuitive Uniform Resource Locator (URL) syntax for making server requests. OPeNDAP represents a file's contents with a Dataset Descriptor Structure (DDS) and Data Attribute Structure (DAS) text document that a client obtains from the server, or with an equivalent DDX XML[1] document. The subset request is made using array indexes, since the OPeNDAP data model, like NetCDF and HDF5, describe its data objects as multidimensional arrays. The reference implementation for OPeNDAP servers is the Hyrax server from opendap.org. The DAP protocol is an open standard that defines an on-the-wire data representation which is implemented by other servers, including DAPPER[2], GrADS-Data Server (GDS), Live Access Server, PyDAP[3], and the Thematic Realtime Environmental Distributed Data Services (THREDDS) Data Server (TDS). Each server has its own set of file formats and data types that it knows how to serve. The DAP protocol allows clients to access them in a uniform, but low-level, way.

---

[1] DDS in XML format.

[2] http://www.epic.noaa.gov/epic/software/dapper/

[3] Python library implementing the Data Access Protocol.

5.2.3 *Remote Subsetting in Coordinate Space*

Making requests in index space implies that a client understands the semantics of the data's coordinate systems, and can translate desired space and time subsets into the equivalent index ranges. This is a non-trivial problem and a barrier for casual use of data. There is a very strong drive to develop protocols in which the request for a data subset can be made in coordinate space, e.g. latitude/longitude and time ranges, with the server understanding the coordinate system and returning the correct subset.

The main focus for developing such protocols is the set of web services being developed under the auspices of the Open Geospatial Consortium (OGC), an industry and government group for geospatial standards that has evolved from Geographical Information System (GIS) vendors and their customers. This has been a slow and sometimes fitful evolution of functionality starting from the traditional GIS "2D image" view of data, towards a more complete multidimensional view needed for scientific data. The most successful and widespread OGC web service is the Web Map Service (WMS), which indeed produces 2D images for web displays and other places where a picture of the data is desired. The Web Feature Service (WFS) produces vector graphics, again reflecting traditional GIS object models. The Web Coverage Service (WCS), in contrast, can return real data arrays in response to client requests that are made in coordinate space, and so is a major focus of attention by the scientific community investigating its potential uses. There are two major issues that WCS continues to struggle with. Scientific data is highly heterogeneous, and generalities in any data model can result in unacceptable inefficiencies in an actual implementation. The OGC/ISO data models for WCS data therefore continue to be debated, as various data types and use cases are considered. The other issue is the format of the returned data. Each community has its own preferred data format, and there is no agreed-upon common format. This leads to the likelihood of incompatible islands of WCS servers and clients, each developed for a particular community, without real interoperability. As of this writing, the WCS specification is rather unstable, and it is unclear which communities will end up adopting it and for what purposes.

There are other efforts for remote subsetting in coordinate space. The OPeNDAP community is investigating extensions to the protocol using "server-side functions" for allowing requests in coordinate space. The Unidata TDS has experimental Representational State Transfer (REST) web services with a subset of WCS functionality which returns data in CF compliant NetCDF files, called NetCDF subset services. Other groups and servers are adding coordinate-space subsetting in ad-hoc or experimental ways. Whatever the request protocol or response format is, the key functionality is the ability of the server to translate coordinate values into the data model of the underlying file, and to efficiently extract and encode the response. This requires that the data server understand not just the file format, i.e. the physical layout of the data, but also the georeferencing semantics implicit or explicit in the data.

5.3 Georeferencing Coordinate Systems and Metadata

Data formats such as NetCDF provide a flexible framework for the representation of earth system datasets. This presents a number of choices to a dataset producer: How should the coordinate system associated with a variable be represented. How should time be encoded? How should the dimensions of a variable be ordered? How should variables be named? To the data user, the fewer ways these questions can be answered the better. It is much easier to write programs to interpret and use data if the choices are constrained. Similarly the producers of earth system datasets want to write data that is readily usable by standard software, provided all the requisite information can be represented. This is the purpose of a metadata convention.

5.3.1 CF Conventions

The CF conventions grew out of efforts to standardize NetCDF datasets generated by general circulation climate models (GCMs). CF is based on the NetCDF model of data, but is applicable to other formats as well. Although the standard is designed for GCM datasets, it can be used in related domains such as numerical weather forecasting. CF development was guided by several principles:

- Data should be self-describing, without reference to external tables.
- Conventions should be developed only as needed, rather than try to anticipate all possible needs.
- Conventions should seek to balance the needs of data consumers and producers.
- Metadata should be both human-readable and machine interpretable.

Climate models operate on discretized regions of space and time, so a key goal of the CF convention is to define how the coordinate systems describing such discretization should be represented. This entails:

- Categorization and description of the various spatial coordinate systems and projections used in earth system models, for example, how to unambiguously define longitude and latitude coordinate variables.
- Representation of time and calendars. Climate models often use nonstandard calendars such as 360 day calendars—twelve months of 30 days each.
- Interpretation of data within a grid cell—for example, if the value is an average over a time interval or is instantaneous.
- Consistent naming of variables with standard names. This allows data users to recognize when data from different files or datasets are comparable.
- Descriptions of the data itself—what are the units? How are missing data values defined? What is the valid range of data?

5.3.2 Other Georeferencing Conventions

Formats such as NetCDF and HDF5 are designed for the general case of scientific data, whose underlying data types are multidimensional arrays of primitive types and compositions of those. Data access APIs have no notion of coordinates, or of data types such as grids, images or points. These semantics are added by a combination of conventions which clarify how data is stored and what certain metadata means, and other libraries and their APIs which implement these conventions and provide convenience routines to the programmer. HDF-EOS is an example of a specialization of HDF4 and HDF5 files for NASA data, consisting of a library that sits on top of the HDF 4/5 library, as well as a set of conventions for how data is stored. The HDF-EOS library has explicit notions of coordinate systems and data types, and programmers using that library are relieved of the low-level details of how these are represented.

Specialized formats include higher levels of semantics in the file format specification itself. For example GRIB, which only stores 2D gridded data, explicitly stores georeferencing coordinate systems in the Grid Definition Section of the GRIB record. The meaning of the data is described in external tables, which are standardized and maintained by the WMO. There are both advantages and disadvantages to external tables and specialized data formats, but GRIB is certainly a widespread and successful example of this approach.

Bespoke and proprietary data formats often omit explicit specification of coordinate systems and other metadata. In using these formats, you "just have to know" certain information in order to be able to use the data. Typically this information is written down as a human-readable document, although perhaps not publically available. This has been a common practice in the past to create "vendor lock-in", that is to ensure that the company that originally wrote the software would obtain subsequent maintenance and development revenues, particularly for large government or business software. This practice has all but disappeared in the world of publically funded scientific data management, as customers are rightly demanding the use of standards, open documentation, and open-source software. In any case, there is a strong movement away from bespoke data formats and towards increasingly powerful open source formats and software such as NetCDF and HDF.

5.3.3 Discovery Versus Use Metadata

Coordinate systems and units of measurement are examples of use metadata, which is metadata needed to correctly use and manipulate the data itself. Another category of metadata is called discovery metadata, which is metadata to assist users to find and understand what data is available. Discovery metadata is typically harvested and indexed by a discovery system such as a digital library. The best example of a discovery system for earth science data is currently NASA's Global Change Master Directory (GCMD) which concentrates on datasets used in global change studies.

There are many standards for discovery metadata, including the generic library-science oriented Dublin Core, GCMD's Directory Interchange Format (DIF) standard, and the federally mandated geosciences-specific Federal Geographic Data Committee (FGDC). Many communities create some form of standards for what metadata is placed in the files that they write. Adding high-quality metadata to datasets is a tedious and often unrewarded task. Creating a uniform standard that all communities follow is likely impossible, although efforts to do so and encourage adoption are worthwhile and have a long-term payoff. Data portals and servers, recognizing these problems, are beginning to provide mechanisms to allow metadata to be added to files without modifying the original files. The THREDDS Data Server, for example, allows the data administrator to add metadata in the configuration, or even directly into the dataset itself using the NetCDF Markup Language (NcML). Third-party annotation, similar to social tagging, could add an important new dimension to solving the problem of discovery metadata.

Standard parameter names constitute both use and discovery metadata. This is the task of mapping specific parameters/variables in a dataset to a controlled vocabulary of standard names. This allows specialized manipulation of the data, such as using a formula to create derived quantities, as well as uniform search by discovery systems. Again, there are a number of standard controlled vocabularies in the atmospheric sciences, including CF standard names, GCMD and GRIB. There are also attempts to make more semantically meaningful networks of concepts, sometimes called ontologies.

5.4 Data Models

A data model is a way to talk about the functionality of a particular data representation or storage format from the user's perspective. A data model describes the objects (nouns) and methods (verbs) that are available for use, as well as the meanings (semantics) of each term in user language. Scientific data models are rather less articulated than, for instance, relational database models, and users tend to focus on the Application Programmer Interface (API) that they use. Models for scientific data have tended to be expressed implicitly in library documentation and in the user's "mental model" of data. However, in a data mode-centric view, an API is a binding of a data model to a specific programming language and implementation library, and a storage format is a way to persist a data model to disk. Creating explicit, abstract data models makes it much easier to compare functionality across libraries and languages.

Unidata's Common Data Model (CDM) is an attempt to explicitly model the semantics of scientific data formats and the scientific data types in common use in the atmospheric/earth sciences. It was originally a melding of the data models of NetCDF, HDF5 and OPeNDAP, as implemented in the NetCDF-Java library. Readers for other file formats (GRIB, BUFR, NEXRAD, etc) were added to the NetCDF-Java library in a way that allows the user to access any of them through an object-oriented, extended NetCDF API. The NetCDF-Java library is thus said to implement a common data model for these formats. The CDM is not original in this effort. The Geospatial Data Abstraction Library (GDAL), for example is a C library specializing in gridded (raster) data for a very large number of file formats. There are many other examples of libraries and protocols intended to achieve similar functionality.

In any scientific data model, one finds several layers of the model, implemented in different software layers or specified in a completely different ways:

1. The *data access* layer, also known as the syntactic layer, handles data reading from the file or network protocol.
2. The *coordinate system* layer identifies the georeferencing coordinates of the data arrays. This may be part of the file specification or be encoded in metadata conventions.
3. The *scientific data type* layer identifies specific types of data, such as grid, swath, or point data, and adds specialized methods for each kind of data, for example allowing data subsetting in coordinate space or to search for data using specific data attributes such as station identifiers for point data.

5.4.1 ISO/OGC Models

The ISO 191xx family of specifications is an ambitious international effort to define abstract models for Geographic Information / Geomatics, defining concepts and abstract reference models for GIS. The Open Geospatial Consortium (OGC) builds on the ISO abstract models with implementation models and web services. These models form the basis for current high-level abstractions of data services for simple uses of scientific data. They continue to evolve to deal with more sophisticated needs, but they will not displace specialized data access and manipulation software tailored to scientists' needs any time soon.

5.5 Data Aggregation

The output of a climate model experiment is typically a collection of 3-D or 4-D variables, where the dimensions of the variables are time, vertical level, latitude, and longitude. Because a climate run may save output for a long period of simulation time—for example, four times daily for a 100 year period—a variable may be represented as a collection of files where each file covers an interval of time. From the end user's perspective it is convenient to maintain the view of a variable as a single array. The process of providing this unified view is called *data aggregation*. There are a variety of tools and services available that support aggregation. We will illustrate two such utilities: the TDS and the Climate Data Analysis Tools (CDAT).

The TDS uses the NetCDF Markup Language (NcML) to define aggregated datasets. The client only sees the aggregation, not the individual data files. TDS supports three types of aggregation:

- Union: The union of all the dimensions, attributes, and variables in multiple NetCDF files.
- JoinExisting: Variables of the same name (in different files) are connected along their existing, outer dimension, called the *aggregation dimension*. A coordinate variable must exist for the dimension.
- JoinNew: Variables of the same name (in different files) are connected along a new outer dimension. Each file becomes one . A new coordinate variable is created for the dimension.

For example, if two NetCDF files contain data for one or more variables for January and February, respectively, the following NcML contained in a TDS catalog would define an aggregation of the two files:

```
<netcdf xmlns="http://www.unidata.ucar.edu/namespaces/netcdf/ncml-2.2">
  <aggregation dimName="time" type="joinExisting">
    <netcdf location="file:/test/temperature/jan.nc" />
    <netcdf location="file:/test/temperature/feb.nc" />
  </aggregation>
</netcdf>
```

The CDAT utility supports aggregation with the *cdscan* utility. In basic usage cdscan takes as arguments the files to be aggregated, and generates XML markup. For example:

```
% cdscan -x test.xml u_*.nc
Scanning files ...
u_2000.nc
Setting reference time units to days since 2000-1-1
u_2001.nc
u_2002.nc
test.xml written
```

Within CDAT the XML file is treated as an ordinary NetCDF file, and appears to contain all three years of data:

```
% cdat
>>> import cdms2
>>> f = cdms2.open('test.xml')
>>> f['time'].asComponentTime()
[2000-1-1 0:0:0.0, 2001-1-1 0:0:0.0, 2002-1-1 0:0:0.0]
```

CDAT can aggregate any data format it supports, such as NetCDF, GRIB1, HDF4, or Met Office Post Processing (PP) format.

5.6 Case Study: Attribution and Detection

This case study is based on an attribution and detection analysis of human-induced changes in atmospheric moisture content [Santer et al. (2007)]. The purpose is to illustrate the various aspects of data representation from the vantage point of a student who is tasked with carrying out the analysis. Computational details of the analysis are omitted. The goal of the study is to determine if there is an anthropogenic signal present in recent observations of atmospheric moisture. The student proposes to carry out a "fingerprint" analysis to determine if observed increases in atmospheric water vapor over the ocean, as measured with the satellite-borne Special Sensor Microwave Imager (SSM/I), can be explained by natural climate noise alone or have a man-made component. She uses model data from the World Climate Research Programme Coupled Model Intercomparison Project phase 3 (WCRP/CMIP3) Multi-model Database (MMD), the archive used for the 2007 Intergovernmental Panel on Climate Change (IPCC) Fourth Assessment Report. "Climate of the 20th century (20c3m)" simulations will be analyzed to see if they capture the observed high-frequency variability of atmospheric water vapor. The MMD "pre-industrial control (picntrl)" experiments will be used to examine whether estimates of internal climate "noise" can explain observed increases in water vapor.

The student's initial task is to locate and assemble the data. From the MMD web portal she finds that the database contains 934 datasets, each representing a climate simulation for one of twelve predefined experiments, including picntrl and 20c3m. Each dataset contains one or more variables, but not all modeling groups have submitted water vapor data. Each dataset has discovery metadata attributes such as:

- Experiment, for example "20c3m"
- Model identifier
- Temporal frequency (3-hourly, daily, monthly, or yearly averages)
- Run identifier, for ensemble runs.

The IPCC 20c3m experiments were run with imposed external forcings that differed between modeling groups. Some experiments included changes in both natural forcings, such as volcanic dust and solar irradiance, and anthropogenic forcings such as carbon dioxide and other greenhouse gases. The analysis will separate the datasets into two groups, one having all types of forcings and the other anthropogenic-only forcings. Unfortunately the MMD standardized metadata does not list the specific external forcings used by each model. The student contacts each modeling group for the additional information, and determines that ten of the models used only anthropogenic forcings, while twelve of the models included changes in both natural and anthropogenic influences.

To find the specific variables needed, the student searches a list of standard names on the MMD web site and finds a CF standard name of "atmosphere_water_vapor_content", associated with the name "prw" (short for "precipitable water"). A search on the website reveals 137 files containing water vapor data for the 20c3m experiment, a total of 5.6 GB; a similar search of the control experiment finds 6.2GB of data. The

MMD provides direct access to the data through an OPeNDAP server, which serves aggregations of the datasets. One option is to process the data directly from the server, since the available analysis tools are compiled with the appropriate OPeNDAP client libraries. However because the computation will make several passes through the data, she chooses to download the files to local disk for improved performance.

The SSM/I observations are over ocean only, so the land-sea mask for each model is also needed. This variable defines the percentage of each grid cell occupied by land. (The CF standard name is "land_area_fraction"). Together with the grid description contained in each file, which defines the area of the cells, the mask information is carried through the computation to ensure a precise comparison of observed and model data.

Now that the data has been assembled in a common form—CF-compliant NetCDF files—the processing phase of the analysis begins. The student notes that each model has produced data on different rectilinear grids. She regrids all the model and observed datasets to a common $10° \times 10°$ grid. There are a number of utilities for regridding—in this case she uses the SCRIP package Jones (1999). SCRIP provides a suite of standard interpolation techniques to choose from, such as bilinear or conservative remapping. Regridding the data will facilitate comparison and decrease the computational load without a significant loss of information, since the water vapor fields vary smoothly. The student uses the subsetting API of NetCDF to extract model data in the latitude range $50°N–50°S$, the approximate range of the observed data. In the remainder of the analysis (which we skip over—see Santer et al. (2007) for the details) she calculates the leading empirical orthogonal functions (EOFs) for each combination of (all forcings, anthropogenic forcings) and (20c3m, picntrl) datasets. Using the EOFs she derives a signal-to-noise function which indicates that the observations do indeed have a statistically significant anthropogenic signal.

References

Jones P (1999a) First- and second-order conservative remapping schemes for grids in spherical coordinates. Mon Weath Rev 127:2204–2210

Santer B, Mears C, Wentz F, Taylor K, Gleckler P, Wigley T, Barnett T, Boyle J, Bruggemann W, Gillett N, Klein S, Meehl G, Nozawa T, Pierce D, Stott P, Washington W, Wehner M (2007) Identification of human-induced changes in atmospheric moisture content. Proc Nat Acad Sci U S A 104:15248–15253

Chapter 6
Data Analysis and Visualization

D. N. Williams, T. J. Phillips, S. C. Hankin and D. Middleton

6.1 Background and History

The Earth's climate—its average weather and frequency of extreme events—greatly affects the conditions of all living creatures. Human conditions, for example, are strongly influenced by the availability of fresh water, the ambient temperature and humidity, and extreme phenomena such as heat and cold waves, droughts and floods, and tornadoes and hurricanes. Dramatic changes in climatic phenomena (e.g. a Medieval warm period or the prehistoric Ice Ages) also have been recorded, either in human chronicles or in the natural history of the Earth itself.

Moreover, the rise of the industrial revolution and the burning of fossil fuels with the release of carbon dioxide and other heat-trapping greenhouse gases have brought about a human impact on climate that augments natural mechanisms which are only partially understood. The post-industrial average global temperature has risen more than one degree Celsius, with increases of several more degrees possible, depending on future fossil fuel emissions. With this increase in global temperature will come a number of potentially dire consequences for living creatures and their habitat.

D. N. Williams · T. J. Phillips
Program for Climate Model Diagnosis and Intercomparison, Lawrence Livermore National Laboratory, 7000 East Ave, Livermore CA 94550, USA
e-mail: williams13@llnl.gov

T. J. Phillips
e-mail: phillips14@llnl.gov

S. C. Hankin
National Oceanic and Atmospheric Administration, Pacific Marine Environmental Laboratory, 7600 Sand Point Way NE, Seattle WA 98115, USA
e-mail: steven.c.hankin@noaa.gov

D. Middleton
National Center for Atmospheric Research, 1850 Table Mesa Dr, Boulder CO 80305, USA
e-mail: don@ucar.edu

V. Balaji et al., *Earth System Modelling – Volume 4*, SpringerBriefs in Earth System Sciences, DOI: 10.1007/978-3-642-36464-8_6, © The Author(s) 2013

It is important to understand the possibly perilous future climatic changes in order to better adapt to such outcomes or to mitigate them as much as possible. To better understand where the Earth's climate might be headed, scientists have developed computer models to quantitatively simulate the future climatic states of the planet's atmosphere, ocean, ice, land, and biosphere.

For many years, our collective understanding of weather and climate was largely qualitative in character; but the advent of computers made possible the detailed quantitative analysis of such processes. However, it was only in the late 1960's that scientists were able to develop a climate model that combined atmospheric and ocean processes in an idealized form. This first "general circulation model" implemented at the National Ocean and Atmospheric Administration (NOAA) Geophysical Fluid Dynamics Laboratory in Princeton, New Jersey signified a pivotal breakthrough that has subsequently been developed to a much higher degree of realism. With further advances in computer technology, Earth systems models that are also able to realistically simulate terrestrial hydrology, interactive vegetation, and biogeochemical cycles are becoming increasingly possible.

In order to interpret the massive amounts of data produced by today's century-scale climate simulations, effective software utilities for data analysis and visualization also must be developed. Because no single computer model is best for simulating all aspects of the climate system, tools and techniques to diagnose and inter-compare the behaviors of different models are also essential.

The process by which scientists evaluate model behavior against available observations is based on the need to understand the simulation data and to readily convince other scientists of their interpretations (Fig. 6.1). Initially, analysis software—developed for specific model output and tasks—ran on the same machines as the models. These primitive FORTRAN routines were rather difficult to use and were even more difficult to share within the scientific community.

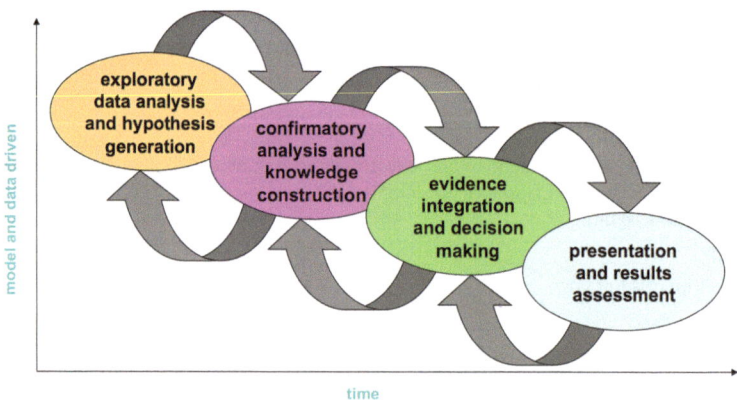

Fig. 6.1 Scientific inquiry of geospatial information–a seamless process for exploring and analyzing climate information

In the 1950's and 1960's, visualization of model results often consisted of merely a single one-dimensional line plot or two-dimensional contour map. As computer technology progressed further in the 1970's, visualization libraries such as the Graphical Kernel System (GKS) were developed, offering consistent and complete one- and two-dimensional graphics standards. Then, with the advent of standard visualization libraries, came the integration of analysis packages (e.g. the FORTRAN coded SPHEREPACK and LINPACK). With the recent development of more powerful workstations and personal computers, the availability and diversity of data analysis and visualization options has exploded, generating many powerful suites of tools applicable to the diagnosis of climate model behaviors. Today we are only beginning to see a convergence of these options, and their increasing interoperability because of the development of standards and best practices.

Today's tools greatly aid climate research by allowing flexible work strategies. Open-source development paradigms and component-based mechanisms permit the integration of legacy code (e.g. the FORTRAN-based SPHEREPACK), improve inter-operability among diverse groups, and promote national and international collaborations. These analysis and visualization tools also are collocated with the data, giving a user the option of remotely processing, rather than inefficiently moving data over long distances for local examination. Increasingly the costs associated with moving large model outputs across the Internet are driving decisions on how and where to deploy analysis and visualization tools. An emerging practice is to deploy the tools "close" (as measured by network capacity) to the data, and to provide the means for users to exploit these tools through remote connections—an approach consistent with the emergence of Service Oriented Architectures.

Data analysis and visualization systems for climate models are on a path to become much more sophisticated services, allowing collaborative research groups to conduct simultaneous analysis and visualization of extreme-scale datasets. As part of a much larger scientific discovery infrastructure, data analysis and visualization sites will exploit multiple processors and cores to enable parallel analysis and visualization wherever possible.

6.2 The Analysis Process

Analysis of model output proceeds in several over-lapping phases: analysis performed by the developer in order to "tune" the model to render a better simulation; validation of a model simulation against observations; analysis to assess the scientific implications of results from an individual simulation relative to those of a similar run; analysis performed by small communities of developers engaged in model inter-comparison experiments; analysis undertaken by diverse specialists such as ecologists or economists who use climate simulation data as inputs for their impacts models.

The analyses associated with tuning the model's behavior generally occur at the modeling centers and begin with "batch" processes that are often bundled into the model execution runtime environment. The software to perform this processing is often model- or framework-specific, frequently using code from the model itself. Calculations and visualizations generated in this mode are usually tailored to the priorities of the developers, such as the need to diagnose long-term drift in the model state variables. (A sobering indication of the preliminary level of development of today's Earth System Models is that tuning to achieve long-term stability often includes a "trial and error" investigation of nonphysical or unstable model behaviors.)

The next analysis phase focuses on assessing the scientific implications of results from individual model runs, most commonly through comparison with similar runs over the same domain and coordinate system. Such comparisons allow the modeler to pinpoint the effects of specific changes to initial conditions, external forcing, or model physics that were introduced in that model run. Broad arrays of desktop software are available to perform this sort of differencing plus visualization, as described in the next section. The analyses will be both synoptic (comparing the same time step or averaged time period from different runs) and time-series oriented. The ability to smoothly blend these two outlooks imposes a flexibility requirement on the data management strategy for model outputs. It is preferable that the user interacts with the full, 4-D (space-time) coordinate reference system of the model output without the encumbrances associated with data distributed across discrete files.

The task of comparing simulations involves the examination of both model state variables from the history files and of diagnostic fields generated by batch analyses. Since it is highly desirable to use the same software tools for both these types of data, greater efficiency is realized by utilizing consistent data management practices for both the output of diagnostic quantities and model outputs. The CF conventions[1] with their associated NetCDF file format and file aggregation techniques are increasingly recognized as suitable standards for both classes of data.

Intercomparison of outputs from an ensemble of distinct models is rapidly becoming a routine requirement for climate data users. (The associated complexities of distributed data management and access control are described in Vol. 6.) A key complexity of such an intercomparison is that each model's output typically will be represented on a different grid—and increasingly on a complex curvilinear or tiled space such as that of the tri-polar[2] or the cube-sphere grid.[3] Thus the analysis and comparison of model outputs requires software algorithms for re-gridding between coordinate systems. Stand-alone utilities such as fregrid[4] and SCRIP[5] are available for this purpose, but transparent access to equivalently advanced re-gridding functionality through desktop analysis and visualization tools is in an early stage of development. Several efforts are underway to embed these re-gridding capabilities into Web Services, but none are operational at this time.

---

[1] http://cf-pcmdi.llnl.gov/

[2] http://nomads.gfdl.noaa.gov/CM2.X/oceangrid.html

[3] http://www.gfdl.noaa.gov/~vb/gridstd/gridstdse2.html

[4] http://www.gfdl.noaa.gov/~vb/grids/gridspec-tools.html

[5] http://climate.lanl.gov/Software/SCRIP/

The comparison of climate model outputs with observations is complicated by the nature of Earth observational data (particularly historical observations), which suffer from inconsistent quality control, insufficient metadata, and spotty coverage. The standards and data templates for describing observations are still emerging, thus the corresponding software systems are immature with severe interoperability problems. Climate data users typically work with gridded data products e.g. the Comprehensive Ocean-Atmosphere Data Set (COADS[6]) and World Ocean Atlas[7] or model-based reanalyses that are derived from observations. Because such datasets are similar in structure to model outputs, the same software tools can also handle them. Observations, such as time series at or vertical profiles at particular locations are similar in structure to some batch diagnostic results, so the consistent use of the CF conventions can also make analysis of these observations tractable. Over the coming years we can expect to see an increasing convergence on standards for the representation and interchange of observations, such as those from the Open Geospatial Consortium [8] and those emerging from Unidata and the Common Data Model developments for NetCDF[9] and CF.

Users from other fields of research, such as economists and ecologists, who wish to include the outputs of climate models in their analyses, most generally want access to the model results through formats or protocols that are compatible with the tools they already use. Such tools may include spreadsheets, GIS applications, and general-purpose mathematical and statistical software packages, such as MATLAB© or S©. Increasingly these needs are being met through Web Services such as those from OGC and Web servers, such as the Live Access Server,[10] that can regrid on demand and provide ASCII, tab-delimited, and GIS-friendly formats for subsets of model outputs.

6.3 Current Platforms Used for Data Analysis and Visualization

The design of climate application software is driven by the need for seamless exploration of multi-dimensional model and observational data (Fig. 6.1) on a plat form that can access an ever-growing selection of analysis approaches. Moreover, the priorities of particular climate research groups–and ever changing computer platforms, operating systems, and software languages–make user education very challenging. As the climate community moves toward developing a "built-to-share" scientific discovery infrastructure (i.e., for the mutual benefit of researchers, policymakers, and the larger society), users will increasingly access elements of the process platform with varying degrees of expertise; but today's platforms require

---

[6] http://icoads.noaa.gov

[7] http://www.nodc.noaa.gov/OC5/WOA05/pr_woa05.html

[8] http://www.opengeospatial.org

[9] http://www.unidata.ucar.edu/software/netcdf

[10] http://ferret.pmel.noaa.gov/Ferret/las

human intervention at every phase of the analysis process—a structure too labor-intensive and limiting for the extreme scale of future computing tasks.

Platforms include many aspects hardware and software. For hardware, the most important component is the visual display that is used not only to browse data, but also to document a simulation's characteristics. Thus, it is necessary that the visualization utilities allow as much user control over the display details as feasible, while also providing hardcopy visuals with minimal human intervention.

Because visualization is a means for interpreting climate data quickly, it is imperative that effective graphical utilities be developed. Multiple graphics methods and visualization techniques also are necessary to assist the scientist to comprehend the details of climate simulations. The corresponding visualization model thus must be able to access different graphical engines that reside close to the data, thereby broadening the user's scope of understanding. Visualization also can play a major role in helping policymakers or other non-specialists to comprehend the basic features of particular climate data. Visual representation and interaction techniques for production, presentation, and large-scale dissemination therefore also must be central to the visualization model.

Platforms must provide the user with the ability to control the process, either interactively or via a script. A scripting capability also allows the instantaneous state of the system or an entire interactive session to be saved for later recovery, replay, or editing. Such features can be realized on specialized platforms, but these demand a high level of operational expertise. Instead, these capabilities need to be implemented on the local workstations, PCs, or laptops that are favored by most users.

6.4 Community Tools and Environments

Climate research scientists require a suite of interrelated diagnostic software tools that are inexpensive, efficient, consistent, easy to use, flexible, portable, adaptable, shareable, and capable of operating in a distributed environment. The climate modeling community is developing a number of open-source community software tools that are well on their way to fulfilling these requirements. These include, for example, the Earth System Grid (ESG), Collaborative Climate Community Data and Processing Grid (C3Grid), Climate Data Analysis Tools (CDAT), Grid Analysis and Display System (GrADS), NCAR Command Language (NCL), Ferret, and the NetCDF Operator (NCO) and Visual Browser (Ncview). (Two popular proprietary tools are the Interactive Data Language (IDL) and MATLAB©, but these are not specifically designed to ingest standard climate model output or to produce specialized visualization displays.)

Existing within larger problem-solving systems, these easy-to-use tools link together disparate software to form an integrated analysis environment. Specialized analysis and visualization software often also is deployed within this tool-enabled environment, and then is distributed to the broader climate community. General numerical packages developed mainly by other scientific communities also are commonly used for implementing diagnostic and statistical applications, as well as other data manipulation functions.

The shared climate analysis packages are compliant with community-endorsed conventions such as the Climate and Forecast (CF) metadata that promote the processing and sharing of files created with the widely used NetCDF application-programming interface. These analysis tools are used to extract climate variable properties, such as dimension, grid, vertical level, and temporal information. Following on analysis tasks, visualization is then implemented by various software packages. The concept of community tools and environment must be flexible enough to allow for expansion and interchange of a future infrastructure.

Over the next several years, today's already considerable capabilities will be expanded to allow dissemination of a much broader set of climate data that will include in-situ and satellite observations, additional multi-model simulations that include dynamical vegetation and biogeochemistry, and ultrahigh-resolution variables (e.g., at 1 km scale) rendered on the Mosaic grids (i.e., an assembly of smaller structured meshes (tiles)) that are increasingly popular. An open-source software system also should encourage contributions of new applications by members of the broad climate community.

Utilizing current and future technologies and computational resources, the community is developing a sophisticated data distribution system that will allow users to remotely access, visualize, and diagnose petabyte-scale data in a secure environment. A notable current example is the multi-model data archive created on behalf of the Intergovernmental Panel on Climate Change (IPCC) Fourth Assessment Report (AR4), now known as the Coupled Model Intercomparison Project 3 (CMIP3) multi-model database. As a component of the larger infrastructure, analysis tool users can manipulate all aspects of CMIP3 datasets. In this complete workflow environment, these users can gain access and direct the flow of data through a script file, stand-alone client, or web-based interface.

Among groups whose primary interests concern the biological and societal impacts of climate change, there is a growing interest in data that extend beyond conventional climate model data products and that are archived in formats readily accessible to Geographic Information System (GIS) applications. For model data to be made more relevant for this broad community of researchers, educators, students, and policy makers, new services aimed at delivering CF-conforming distributed data in GIS format are therefore needed.

In order to prevent unauthorized data access and system use, creation of security safeguards also will be necessary at every level of a "built-to-share" scientific discovery infrastructure, for example to handle service-level authentication and message encryption and integrity. These requirements can be implemented via an interface that allows users to register and be affiliated with a group that is entitled to access specific logical resources such as high-speed data transfer, data analysis, and visualization.

Climate model datasets are growing at a faster pace than those for any other field of science. Given current growth rates, these datasets will collectively amount to hundreds of exabytes by the year 2020. To provide the international climate community with convenient access and to maximize scientific productivity, these data should be replicated and cached at multiple locations around the globe. Unfortunately, establishing and managing a distributed data system poses several significant challenges, both for data analysis and visualization development, and for the existing wide area and campus networking infrastructures. For example, transport technologies currently deployed in wide area networks do not cost-effectively scale to meet the scientific community's projected aggregate capacity needs, based on the projected dataset growth rates. Even if backbone network technology improvements increase link speeds tenfold from the current 10 GB/s (as anticipated in the next several years), more efficient networking resources will be essential. Efforts are now underway to develop hybrid networks with dynamic virtual circuit capabilities, such as those currently being tested and deployed by the Energy Sciences Network (ESnet). Although dynamic virtual circuits allow high-capacity links between storage and computer facilities to be created as needed (and then to be quickly deactivated to free up network capacity for other users), much work must still be done to optimize and harden the relevant software.

6.5 Use Case Example

The following use case illustrates the interface of the researcher/user with the data in order either to perform scientific research or to understand environmental concerns relevant for setting policy. The processing effort involves moving vast amounts of data (spanning several government agency analysis centers) to and from various sites around the world.

In this example, domain experts (i.e. climate scientists) in different international locations are called upon to provide crucial inputs for policy decisions on outbreaks of African malaria that depend on local climate conditions. From their remote sites, the scientists search a climate portal containing exabytes of high-resolution regional central Africa data. When their data search proves unproductive, the scientists run several models in real-time to generate ensemble simulations of African climate. Using server-side visualization tools, they are able to simultaneously view and annotate plots of ensemble climate statistics on their respective platforms. The climate scientists then save this session, and later a malaria policymaker discovers (using a "new search capability") the provenance of this saved session. Working with the scientists, the policymaker conducts further assessment and re-analysis of the derived datasets before reducing to 20 TB (from their original size of 20 PB) and moving them to a local workstation for further study. The scientists then integrate the climate model ensemble data with the African malaria data for potential future use.

Thus, although the targeted primary users are domain experts, it is essential that non-experts (e.g., politicians, decision makers, health officials, etc.) also are able to access much of the data. Rich but simple interfaces will allow the non-experts to accomplish difficult data manipulation and visualization tasks without having to understand the complexities of application programs or the computing environments on which they depend (Fig. 6.2).

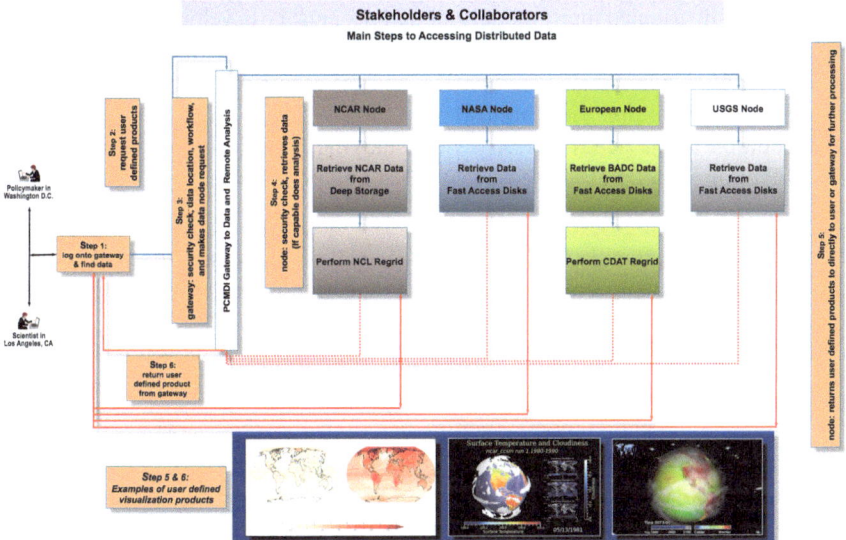

Fig. 6.2 Schematic depiction of a use-case scenario supporting remote login for experts (e.g., model developers, climate researchers) and non-experts needing fault-tolerant end-to-end system integration and large data movement. For the expert user, the system expresses all the capabilities and complexities needed for rich data exploration and manipulation. For the non-expert, however, a simple abstraction layer includes easy-to-use controls for maneuvering within the system

6.6 Future Challenges and Directions

As is now apparent, simulation data are as intricate as the models that produce them. Exhibiting much of the complexity of the Earth's weather and climate, model data can be rendered in many different compression formats, conventions, and images. Climate data also include environmental information of other types that are obtained from in situ and remote observations or derived from the Earth's historical geological and biophysical records. In all cases, these data symbolize the inter-relationships of multiple facets of the nonlinear climate system.

Initially produced at modeling centers housing supercomputers and peta-scale archival storage systems, large-scale climate model data often are translated into forms more convenient to move or to process. In today's distributed computing world, the data discovery, access, and analysis tasks can be challenging, and thus services and tools are developed in order to support users' end-to-end data processing needs. These include data publication services, user registration and management, policy-driven access protocols, local and remote analysis capabilities, and system monitoring and user metrics feedback–all to make distributed climate simulation data easily accessible.

To more reliably predict future weather and climate, multi-model simulation ensembles are analyzed. In order to support this user community, an infrastructure for accessing distributed multi-model simulations must be developed. The necessary capabilities include tools for model variable identification and intercomparison, for application of standard statistical metrics, and for access to observational datasets for validation of multiple simulations.

In addition to the needs of researchers studying the science of climate change (IPCC Working Group I), those of two other important professional communities should not be overlooked:

- The impacts community (IPCC Working Group II) that assesses the scientific, technical, environmental, economic and social aspects of the vulnerability (sensitivity and adaptability) to climate change; and
- The policy community (IPCC Working Group III) that is concerned with the economic, political, and social aspects of climate-change adaptation/mitigation strategies.

The Working Group I community typically utilizes a variety of analysis and visualization tools such as CDAT, NCL, and Ferret. These applications require that the user have an intimate knowledge of the data and of the computational environment where they are situated. Acquiring such detailed knowledge may be impractical for the Working Group II and III users, however. For these communities, simpler Geographic Information System (GIS) capabilities are needed to effectively access and utilize rich data holdings, thereby fully leveraging the substantial modeling investment.

In conclusion, many scientists now are embracing the collaborative opportunities and benefits of emerging tools and environments (e.g. Virtual Globes, GoogleEarth, Science on a Sphere, server-side processing, collaborative visualization, etc.). These tools will need to allow access to climate data by those with little knowledge of the archival or processing environment. Only then will large professional communities (i.e., scientists, researchers, students, educators and policymakers) have the resources necessary to significantly expand our collective understanding of climate change and its environmental and societal impacts.

Chapter 7
Future Perspectives

V. Balaji

In order to summarize the advances in I/O technology outlined in this Brief, and chart a course to the future, it is necessary to appreciate how far we have come by looking back into the past, and the early years of supercomputing. In the pioneering era of Seymour Cray (note his epigram at the start of the Brief) I/O was somewhat of an afterthought. The Cray vector machines introduced the Flexible Formatted I/O (FFIO) software layer that enabled setting and configuration of buffers and read/write layers which could be directly optimized for the disk hardware layer underneath. It required detailed knowledge of spindle and sector layout and exposed them directly in scientific code. Any changes to the underlying hardware meant rewriting considerable amounts of I/O code. The data models—ways to understand and interpret the actual byte sequences—were also hardware-specific and subject to change between systems.

Chapter 2 shows how far we have evolved since those early days. Though the parallel applications are vastly more complex, and the underlying disk subsystems equally evolved in sophistication, a fairly stable layer of abstraction is now commonplace, and shields application developers from the gory details of disk subsystems. While there are still some issues that can interfere with optimal I/O performance, such as the "false sharing" problem described in Sect. 2.2, it is relatively easy to diagnose and correct for the majority of bottlenecks in I/O optimization. Chapter 4 addresses the auxiliary issues that arise when the data is on linear media such as tape, which does not lend itself to random access. There again, some fairly straightforward optimizations such as the HSM caching layers described in Sect. 4.2, allow us to deal with those complications without delving too deeply into the specific hardware configurations underneath.

Chapter 3 takes this a step further. Abstract classes now allow users to think in terms of *distributed fields*, an intuitive way for scientists to conceive of physical fields under parallel domain decomposition. These classes also introduce metadata

V. Balaji
Princeton University,Princeton, USA
e-mail: balaji@princeton.edu

V. Balaji et al., *Earth System Modelling – Volume 4*, SpringerBriefs in Earth System Sciences, DOI: 10.1007/978-3-642-36464-8_7, © The Author(s) 2013

describing the physical content of the distributed arrays, adding a layer of *semantic* information, where the meaning of the data is also present. We have also made considerable progress in standardizing scientific data file formats such as NetCDF, as described in Chap. 5. Binary data is no longer a cryptic private code accessible only to its creator: it can be placed in public archives where a naive user can examine the contents of a file, and can directly interpret it in terms of physical information, rather than 1s and 0s.

The ability to include semantic information alongside data has enabled a major transformation of the field of Earth System modeling itself. It is now possible to analyze ESM information from multiple sources—observations, different models— on a common footing. This advance, described in Chap. 6, has transformed the science, but has vastly enlarged the scope of the I/O problem. It is not any longer just an issue of input and output under the control of a single application: the data is now a global resource, and subject to unpredictable use patterns, requests that appear and disappear without warning. The caching problem is now even more acute as the data may be geographically distributed. The associated data volumes (Overpeck et al. 2011) also threaten to overwhelm the network and storage hardware that support this vast enterprise.

Together, the chapters in this Brief outline the transformations that have taken place and point to directions for the future. The hardware has been abstracted away to large degree; physical variable now include semantic information encoded into standard data models have made ESM data accessible to a wider range of users; this accessibility has transformed ESM science into a coordinated global enterprise.

We anticipate this trend to continue into the next decade. Storage technologies continue to become more exotic (see for an interesting survey of hardware futures Kogge et al. 2008 for an interesting survey of hardware futures) and will be fit under widely-used parallel I/O abstractions. Globally federated data archives will evolve into adaptive systems for managing resources, intelligently handling replication and caching, and with inbuilt layers of fault-resilience.

There are new challenges emerging in both hardware and software as well. As outlined in Kogge et al. (2008), hardware is trending toward greater heterogeneity, and processing elements may no longer be capable of traditional I/O. Abstractions may have to be mapped on to specialized I/O hardware, as outlined in Chap. 3 (see in particular the discussion of Fig. 3.1). In another intriguing trend, the conventional separation between "memory" and "storage" may soon end, replaced with storage of varying degrees of persistence shared between different classes of nodes in a heterogeneous system. These "storage-class memories" (see e.g., Lam 2010) will involve further programming challenges under the abstractions of high-level I/O libraries.

System software for such large distributed systems is also likely to be quite different. Distributed filesystems in common use now in cloud applications have spawned entirely new software approaches, such as MapReduce (Dean and Ghemawat 2008), and adapting those to the needs of scientific computation is an emerging area of research.

These new approaches show that I/O will continue to be a challenging field of research as we approach the exascale era. Seymour Cray's epigram will certainly still hold true a decade from now and I/O "will still need work", but this survey shows that I/O has transformed itself, and Earth System Science, into a quite different enterprise from anything that was then imagined.

References

Dean J, Ghemawat S (2008) MapReduce: simplified data processing on large clusters. Commun ACM 51(1):107–113

Kogge P, Bergman K, Borkar S, Campbell D, Carson W, Dally W, Denneau M, Franzon P, Harrod W, Hill K, et al. (2008) Exascale computing study: technology challenges in achieving exascale systems. DARPA Information Processing Techniques, Office 705

Lam C (2010) Storage class memory. In: 10th IEEE International Conference on solid-state and integrated circuit technology (ICSICT), pp 1080–1083

Overpeck J, Meehl G, Bony S, Easterling D (2011) Climate data challenges in the 21st century. Science 331(6018):700–702

Glossary

API	Application Programming Interface
AR5	IPCC Assessment Report 4
ASCII	American Standard Code for Information Interchange
BUFR	Binary Universal Form for the Representation of meteorological data
C3Grid	Collaborative Climate Community Data and Processing Grid
CDAT	Climate Data Analysis Tools
CDM	Common Data Model
CDO	Climate Data Operator
CF	Climate and Forecast
CMOR	Climate Model Output Rewriter
DAP	Data Access Protocol
DAPPER	OPeNDAP Server for in-situ and gridded data
DAS	Data Attribute Structure
DDS	Dataset Descriptor Structure
DDX	Dataset Descriptor Structure (DDS) in Extensible Markup Language (XML) format
DIF	Directory Interchange Format
ECMWF	European Centre for Medium-Range Weather Forecasts
ESG	Earth System Grid
ESM	Earth System Model
ESMF	Earth System Modeling Framework
ESnet	Energy Sciences Network
FFIO	Flexible Formatted I/O
FGDC	Federal Geographic Data Committee
FMS	Flexible Modeling System
FTP	File Transfer Protocol
GCDM	Global Change Master Directory
GCM	General Circulation Model
GDAL	Geospatial Data Abstraction Library
GDS	Grid Analysis and Display System (GrADS)-Data Server
GIS	Geographical Information System
GMAO	Global Modeling and Assimilation Office

GPFS	General Parallel File System
GrADS	Grid Analysis and Display System
GRIB	GRIdded Binary
GridFTP	Grid File Transfer Protocol
HDF	Hierarchical Data Format
HDF—EOS	Hierarchical Data Format—Earth Observing System
HPC	High Performance Computing
HPSS	High Performance Storage System
HSM	Hierarchical Storage Management
HTTP	Hypertext Transfer Protocol
IDL	Interactive Data Language
IEEE	Institute of Electrical and Electronics Engineers
IPCC	Intergovernmental Panel on Climate Change
JPEG	Joint Photographic Experts Group
MMD	Multi-model Database
MPI	Message Passing Interface
NASA	National Aeronautics and Space Administration
NCEP	National Centers for Environmental Prediction
NCL	National Center for Atmospheric Research (NCAR) Command Language
NcML	NetCDF Markup Language
NCO	NetCDF Operator
NetCDF	Network Common Data Form
NEXRAD	Next-Generation Radar
NOAA	National Oceanic and Atmospheric Administration
NWP	Numerical weather prediction
OGC	Open Geospatial Consortium
OPeNDAP	Open-source Project for a Network Data Access Protocol
PFS	Parallel File System
PNG	Portable Network Graphics
POSIX	Portable Operating System Interface
PP—format	Post Processing Format, a proprietary file format for meteorological data developed by the UK Met Office
PRISM	Partnership for Research Infrastructure in Earth System Modeling
PVFS	Parallel Virtual File System
PyDAP	Python library implementing the Data Access Protocol
RDBMS	relational database management system
REST	Representational State Transfer
SAM—QFS	Storage and Archive-Manager Quick File System
SCRIP	Spherical Coordinate Remapping and Interpolation Package
SSM/I	Special Sensor Microwave Imager
TDS	Thematic Realtime Environmental Distributed Data Services (THREDDS) Data Server
THREDDS	Thematic Realtime Environmental Distributed Data Services

URL	Uniform Resource Locator
WAN	Wide Area Network
WCRP/CMIP3	World Climate Research Programme Coupled Model Intercomparison Project phase 3
WCS	Web Coverage Service
WFS	Web Feature Service
WMO	World Meteorological Organization
WMS	Web Map Service
XML	Extensible Markup Language

Index